# Stakeholder Perceptions of Utility Role in Environmental Leadership

## About the Awwa Research Foundation

The Awwa Research Foundation (AwwaRF) is a member-supported, international, nonprofit organization that sponsors research to enable water utilities, public health agencies, and other professionals to provide safe and affordable drinking water to consumers.

The Foundation's mission is to advance the science of water to improve the quality of life. To achieve this mission, the Foundation sponsors studies on all aspects of drinking water, including supply and resources, treatment, monitoring and analysis, distribution, management, and health effects. Funding for research is provided primarily by subscription payments from approximately 1,000 utilities, consulting firms, and manufacturers in North America and abroad. Additional funding comes from collaborative partnerships with other national and international organizations, allowing for resources to be leveraged, expertise to be shared, and broad-based knowledge to be developed and disseminated. Government funding serves as a third source of research dollars.

From its headquarters in Denver, Colorado, the Foundation's staff directs and supports the efforts of more than 800 volunteers who serve on the board of trustees and various committees. These volunteers represent many facets of the water industry, and contribute their expertise to select and monitor research studies that benefit the entire drinking water community.

The results of research are disseminated through a number of channels, including reports, the Web site, conferences, and periodicals.

For subscribers, the Foundation serves as a cooperative program in which water suppliers unite to pool their resources. By applying Foundation research findings, these water suppliers can save substantial costs and stay on the leading edge of drinking water science and technology. Since its inception, AwwaRF has supplied the water community with more than $300 million in applied research.

More information about the Foundation and how to become a subscriber is available on the Web at **www.awwarf.org**.

# Stakeholder Perceptions of Utility Role in Environmental Leadership

Prepared by:
**Chris Tatham, Robert Cicerone**, and **Elaine Tatham**
ETC Institute
725 W. Frontier Circle, Olathe, KS 66061

Sponsored by:
**Awwa Research Foundation**
6666 West Quincy Avenue, Denver, CO 80235-3098

Published by:

**DISCLAIMER**

This study was funded by the Awwa Research Foundation (AwwaRF). AwwaRF assumes no responsibility for the content of the research study reported in this publication or for the opinions or statements of fact expressed in the report. The mention of trade names for commercial products does not represent or imply the approval or endorsement of AwwaRF. This report is presented solely for informational purposes.

Copyright © 2006
by Awwa Research Foundation

All Rights Reserved

Printed in the U.S.A.

 Printed on recycled paper

Library
University of Texas
at San Antonio

# CONTENTS

LIST OF FIGURES ................................................................................................ ix

FOREWORD ........................................................................................................... xi

ACKNOWLEDGMENTS ...................................................................................... xiii

EXECUTIVE SUMMARY ..................................................................................... xv

CHAPTER 1: WHAT IS ENVIRONMENTAL LEADERSHIP? ........................... 1
    Overview ......................................................................................................... 1
    How Stakeholders Define Environmental Leadership ................................... 1
    How Water Utility Managers Define Environmental Leadership .................. 2
    Fundamental Characteristics of Environmental Leadership .......................... 3
        1. Environmental Stewardship ................................................................. 3
        2. Collaboration ........................................................................................ 3
        3. Visibility .............................................................................................. 4
        4. Willingness to Be First ........................................................................ 4
    Summary ......................................................................................................... 5

CHAPTER 2: HOW ENVIRONMENTAL STAKEHOLDERS AND WATER UTILITY MANAGERS VIEW THE ROLE OF A WATER UTILITY IN ENVIRONMENTAL LEADERSHIP ......................................................................................................... 7
    Overview ......................................................................................................... 7
    Benefits of Environmental Leadership for a Drinking Water Utility ............. 7
        Benefits According to Stakeholders ........................................................ 7
        Benefits According to Water Utility Managers ..................................... 8
    Reasons Drinking Water Utilities Are Not More Involved in
        Environmental Leadership ....................................................................... 8
    Reasons According to Stakeholders ............................................................... 8
    Reasons According to Water Utility Managers ............................................. 8
    Other Findings from Interviews with Drinking Water Utility Managers ...... 9
    Environmental Leadership Activities that Stakeholders Thought
        Water Utilities Should be Doing ............................................................. 10
    Types of Environmental Leadership Activities that
        Water Utility Managers Indicated They are Doing ................................ 10
    Comparative Analysis of the Main Findings from Interviews with
        Stakeholders and Water Utility Managers ............................................. 11
    Summary ......................................................................................................... 12

CHAPTER 3: HOW CUSTOMERS PERCEIVE ENVIRONMENTAL LEADERSHIP ......................................................................................................... 13
    Overview ......................................................................................................... 13
    General Findings ............................................................................................ 13

    Awareness About Environmental Issues .................................................. 13
    Importance of Water Utilities Being Environmental Leaders ................. 14
    Overall Attitude of Communities Toward the Environment .................... 14
    Is Your Water Utility an Environmental Leader....................................... 14
Perceived Importance of Water Utility Functions ............................................. 15
    Most Important Functions of the Water Utility ....................................... 15
Importance Placed on Environmental Leadership Activities............................. 16
    Importance of Cooperating with Other Organizations to Protect Water . 16
    Importance of Anticipating Impact of Future Development on Water.... 16
    Importance of Encouraging People to Protect Water Sources................. 16
    Importance of Informing the Public to Protect Water Sources................ 17
    Importance of Cooperating with Other Organizations to Protect ALL
        Water............................................................................................... 17
    Importance of Encouraging Users in the Region to Conserve Water...... 17
    Importance of Using Environmentally Friendly Technology.................. 17
    Importance of Helping Fund Protection of the Environment .................. 17
    Importance of Acquiring Land Near Lakes to Protect Water
        Sources From Contamination ....................................................... 17
    Importance of Cooperating with Other Environmental Organizations.... 18
    Importance of Managing All Forms of Water in an Integrated Manner.. 18
    Most Important Environmental Leadership Activities ............................ 18
Overall Satisfaction with Environmental Leadership Issues ............................. 18
    Satisfaction with Water Utility's Cooperation with Others..................... 18
    Satisfaction with Water Utility's Proactive Protection of Drinking
        Water............................................................................................... 18
    Satisfaction with Water Utility's Education Efforts ................................ 19
    Satisfaction with How Often Water Utility Seeks the Ideas of
        Residents........................................................................................ 19
    Overall Satisfaction with Water Supplier ................................................ 19
Willingness to Pay ............................................................................................. 19
    Willingness to Pay More If Water Utility Is an Environmental Leader .. 19
    How Much Respondents Would Be Willing to Pay for Environmental
        Leadership..................................................................................... 19
Support for Environmental Leadership Initiatives............................................. 20
    Support for Various Environmental Leadership Initiatives..................... 20
    Support for Investing in Water Treatment Equipment ............................ 20
    Support for Encouraging Local Governments to Adopt Codes
        For Developers............................................................................... 21
    Support for Working with Other Organizations to Restrict
        Development ................................................................................. 21
    Support for Spending Money on Water Protection Education ................ 21
    Support for Spending Money on Environmental Protection Education .. 21
    Support for Spending Money on Water Conservation Education .......... 21
    Support for Restricting Times and Days for Watering Lawns and
        Filling Pools.................................................................................. 22
    Support for Giving Water Utilities Authority to Restrict Development.. 22

Summary .................................................................................................... 22

## CHAPTER 4: ENGAGING EMPLOYEES TO PROMOTE ENVIRONMENTAL LEADERSHIP ................................................................ 25
Overview .................................................................................................... 25
Examples of this Strategy ........................................................................... 25
    Integrating Environmental Leadership into Organizational Mission Statements and Strategic Plans ............................................... 25
    Using Performance Incentives for Complying with Environmental Regulations ............................................................................................. 26
    Establishing an Environmental Management Position ........................ 26
    Modifying the Application/Hiring Process for New Positions ............. 26
    Distributing Publications on Environmental Leadership Topics .......... 26
    Inviting Employees to Public Hearings on Environmental Issues ........ 27
    Encouraging Energy Conservation and Recycling .............................. 27
How to Prepare an Organization to Implement this Strategy ..................... 27
How to Implement this Strategy at the Work Unit Level ........................... 29
How to Implement this Strategy at the Level of Individual Employees .... 29
Evaluating the Implementation of this Strategy at the Organizational Level ...... 30
Summary .................................................................................................... 31

## CHAPTER 5: BUILDING PARTNERSHIPS TO PROMOTE ENVIRONMENTAL LEADERSHIP ................................................................ 33
Overview .................................................................................................... 33
Examples of Building Partnerships <u>Inside</u> a Service Area ......................... 33
    Cooperative Land Use Planning ........................................................... 33
    Partnerships with Other Governmental Agencies ................................ 34
    Partnerships with Schools .................................................................... 34
    Stakeholder Participation in the Development of the CCR .................. 34
Examples of Building Partnerships <u>Outside</u> a Service Area ....................... 35
    Joint Management of Watersheds ......................................................... 35
    Partnerships with Agricultural Interests ............................................... 35
    Regional Cooperation for Specific Projects .......................................... 35
    Sustaining Adequate Supplies of Drinking Water ................................ 36
How to Prepare an Organization to Implement this Strategy ..................... 37
How to Implement this Strategy ................................................................. 37
Evaluating the Implementation of this Strategy at the Organization Level ......... 39
Summary .................................................................................................... 39

## CHAPTER 6: USING COMMUNICATION TO CHANGE CUSTOMER PERCEPTIONS ABOUT A WATER UTILITY'S ROLE IN ENVIRONMENTAL LEADERSHIP ....... 41
Overview .................................................................................................... 41
Examples of This Strategy .......................................................................... 42
    On-Going Communications .................................................................. 42
    Communicating Directly with Customers ............................................ 42
    Communicating Through Community Organizations .......................... 43

  Combining Communicating Through News Media, Direct
    Communication with Customers, and Communicating
    Through Community Organizations .......................................... 44
 How to Prepare to Implement this Strategy ....................................................... 44
 How to Implement this Strategy ......................................................................... 46
  Steps to Communicate Through the News Media ................................ 47
  Steps to Communicate Directly with Customers .................................. 47
  Steps to Communicate Through Community Organizations ................ 47
 How to Evaluate the Implementation of this Strategy ........................................ 48
 Summary .............................................................................................................. 49

**CHAPTER 7: BUILDING SUPPORT FOR ENVIRONMENTAL LEADERSHIP INITIATIVES** ................................................................................................................ 51
 Overview .............................................................................................................. 51
 Engaging Employees to Promote Environmental Leadership ............................. 51
 Building Partnerships ........................................................................................... 52
 Communication and Investment .......................................................................... 54
  Investing in "Green" Facilities ............................................................... 55
 Impact of the City of Olathe's Efforts ................................................................. 56
  Finding 1: The number of residents in the city who thought it was
   important for the City to be an environmental leader
   increased significantly. ............................................................... 56
  Finding 2: The percentage of residents who thought the City
   was an environmental leader increased dramatically in
   just 13 months. ........................................................................... 57
  Finding 3: The percentage of residents who were informed
   about environmental issues increased significantly ................... 58
  Finding 4: Overall Satisfaction with Fees and Prices Reach New
   Highs. .......................................................................................... 58
 Conclusions .......................................................................................................... 58

**APPENDICES**
 APPENDIX A: Organizing an Environmental Leadership Advisory Board ...... 61

 APPENDIX B: Environmental Leadership Forum Summaries ......................... 65

 APPENDIX C: How to Plan an Environmental Leadership Forum
  in Your Community ..................................................................... 83

 APPENDIX D: Survey Results ............................................................................ 93
  Sample Survey Instrument ..................................................................... 139

 REFERENCES .................................................................................................. 145

 ABBREVIATIONS ........................................................................................... 147

# FIGURES

| | | |
|---|---|---|
| 2.1 | Gaps in Perceptions About Environmental Leadership | 11 |
| 3.1 | How Well Informed Residents are About Environmental Issues | 13 |
| 3.2 | Importance of Water Utilities Being Environmental Leaders | 14 |
| 3.3 | Most Important Services for Water Utilities to Provide | 16 |
| 3.4 | Willingness to Pay | 20 |
| 3.5 | Support for Various Environmental Leadership Activities | 22 |
| 3.6 | Survey Conclusions | 23 |
| 7.1 | Engaging Employees | 51 |
| 7.2 | Building Partnerships | 52 |
| 7.3 | Communication | 55 |
| 7.4 | Implementation Timeline | 56 |
| 7.5 | Importance of Environmental Leadership | 57 |
| 7.6 | Perceptions of Environmental Leadership | 57 |
| 7.7 | Awareness of Environmental Issues | 58 |

# FOREWORD

The Awwa Research Foundation is a nonprofit corporation that is dedicated to the implementation of a research effort to help utilities respond to regulatory requirements and traditional high-priority concerns of the industry. The research agenda is developed through a process of consultation with subscribers and drinking water professionals. Under the umbrella of a Strategic Research Plan, the Research Advisory Council prioritizes the suggested projects based upon current and future needs, applicability, and past work; the recommendations are forwarded to the Board of Trustees for final selection. The foundation also sponsors research through the unsolicited proposal process; the Collaborative Research, Research Applications, and Tailored Collaboration programs; and various joint research efforts with organizations such as the U.S. Environmental Protection Agency, the U.S. Bureau of Reclamation, and the Association of California Water Agencies.

This publication is a result of one of these sponsored studies, and it is hoped that its findings will be applied in communities throughout the world. The following report serves not only as a means of communicating the results of the water industry's centralized research program but also as a tool to enlist the further support of the nonmember utilities and individuals.

Projects are managed closely from their inception to the final report by the foundation's staff and a large cadre of volunteers who willingly contribute their time and expertise. The foundation serves a planning and management function and awards contracts to other institutions such as water utilities, universities, and engineering firms. The funding for this research effort comes primarily from the Subscription Program, through which water utilities subscribe to the research program and make an annual payment proportionate to the volume of water they deliver and consultants and manufacturers subscribe based on their annual billings. The program offers a cost-effective and fair method for funding research in the public interest.

A broad spectrum of water supply issues is addressed by the foundation's research agenda: resources, treatment and operations, distribution and storage, water quality and analysis, toxicology, economics, and management. The ultimate purpose of the coordinated effort is to assist water suppliers to provide the highest possible quality of water economically and reliably. The true benefits are realized when the results are implemented at the utility level. The foundation's trustees are pleased to offer this publication as a contribution toward that end.

Environmental issues are becoming increasingly important to water utilities. This report will help water utility managers better understand what "environmental leadership" means, the benefits of environmental leadership, and how to implement strategies to become an "environmental leader."

Walter J. Bishop, P.E.
Chair, Board of Trustees
Awwa Research Foundation

Robert C. Renner, P.E.
Executive Director
Awwa Research Foundation

# ACKNOWLEDGMENTS

The research team wishes to express sincere appreciation to the four utilities and their staff who participated in the study:

- Kansas City Water Services Department (Kansas City, Missouri)
- City of Olathe, Municipal Services Department (Olathe, Kansas)
- City of Fort Lauderdale, Florida
- City of San Diego Water Department (San Diego, California)

We are especially grateful for the efforts of Frank Pogge, Colleen Newman, and Mabel Ramey-Moore at Kansas City Water Services Department in Kansas City, Missouri; Bill Ramsey and Don Siefert at the Municipal Services Department in Olathe, Kansas; Charles Yackly at the City of San Diego Water Department; and Greg Kisela and Mike Bailey at the City of Fort Lauderdale.

In addition, the authors of this report are grateful for the guidance provided by the members of the Project Advisory Committee - including Sara Katz, Katz and Associates; Dr. Pankaj Parekh, Department of Water and Power, City of Los Angeles; and David Katz, Philadelphia Water Department, and the Awwa Research Foundation Project Manager, India Williams.

# EXECUTIVE SUMMARY

## RESEARCH OBJECTIVES

The primary purpose of AwwaRF Project 2854, Stakeholder Perceptions of Utility Role in Environmental Leadership, was to help water utility managers better understand the concept of environmental leadership and to describe strategies that drinking water utility managers can use to evolve their utilities into environmental leaders. The key objectives of this project included the following:

- Identify the fundamental characteristics of 'environmental leadership' based on input from drinking water utility managers and environmental stakeholders.
- Identify examples of environmental leadership by drinking water utilities.
- Assess the views of drinking water utility managers and environmental stakeholders about the role a drinking water utility should play in environmental leadership.
- Interview residential customers of drinking water utilities to determine their views on the role a drinking water utility should have in environmental leadership as well as related issues.
- Solicit from environmental stakeholders and water utility managers their suggestions for strategies that water utility managers can use to evolve their utilities into environmental leaders.
- Describe basic strategies for environmental leadership in sufficient detail that the managers of drinking water utilities can easily modify the strategies as appropriate, then apply them to their own utilities.

## MAJOR FINDINGS AND CONCLUSIONS

Although this report presents many findings and conclusions, the research team believes the following should be emphasized.

- **The four fundamental characteristics of environmental leadership are: (1) collaboration, (2) environmental stewardship, (3) visibility, and (4) a willingness to be first.** Being perceived as an environmental leader means more that just taking care of the environment. In order to be perceived as a leader, water utilities must be willing to take the initiative to work with other organizations in a visible way that allows their customers, community leaders, and special interest groups to see that the utility is truly committed to the protection and preservation of environmental resources.

- **Residents Think Environmental Leadership Is Important.** The level of importance that residents place on environmental leadership is generally high across the United States. The level of importance respondents from the national survey placed on a water utility being an environmental leader was 91%, based upon the combined percentage of "very important" and "somewhat important" responses. Nationally, residents were 21 times more likely to think it was "very important" for their water utility to be an environmental leader than they were to think it was "not important."

- **In order for water utilities to become environmental leaders, water utility managers will need to change their understanding of what it means to be an environmental leader.** Although water utility managers and environmental stakeholders generally agree about the importance of environmental leadership, there is a major gap in the perceived level of environmental leadership being undertaken by water utiltiies. More than nine in ten of the water utility managers interviewed thought their utility was engaged in environmental leadership activities while less than one in ten of the stakeholders interviewed felt that way. This gap appears to stem from a difference in the way water utility managers perceive their role as environmental leaders compared to the expectations that stakeholder groups have of the industry.

- **Utilities that make environmental leadership a priority before they have a crisis are much more likely to achieve positive results with less effort than utilities that wait.** By being proactive on environmental issues, water utilities are also likely to generate "good will" in the community which will make it easier for the utility to collaborate with environmental interests in the future. This "good will" will help the utility endure crisis in public opinion when an environmental problem occurs because the general public is more likely to believe the water utility has done everything it could to prevent the problem.

- **Many Residents Are Willing to Support Investments in Environmental Leadership.** Forty-two percent (42%) of the respondents from the nation-wide survey indicated that they would be willing to pay up to $2 a month if their water utility was recognized as a leader in protecting and preserving sources of drinking water. Although a majority of those surveyed were not willing to pay for environmental leadership, those who thought their water utility was an environmental leaders (62%) were significantly more willing to pay for investments to protect the environment than those who did not think their utility was an environmental leader (28%). This suggests that one of the most important benefit to a water utility of being perceived as an environmental leader is that customers will be more willing to pay for environmental initiatives.

## PARTICIPATING UTILITIES

Four drinking water utilities partnered with the research team on this project.

- Water Department, City of Fort Lauderdale, Florida.
- Water Services Department, Kansas City, Missouri.
- Municipal Services Department, City of Olathe, Kansas.
- Water Department, City of San Diego, California.
The utilities were chosen to represent different geographical regions of the country. Two

utilities (Fort Lauderdale and San Diego) were chosen to represent water utilities facing the challenges created by rapid population growth, demand to develop land near water sources, and a lack of new sources of drinking water. The other two utilities (Kansas City, Olathe) were selected to represent water utilities in areas where existing water sources are projected to be adequate to meet the growth in demand for drinking water far into the future.

All four utilities are public organizations. Each of these utilities provides water, stormwater and wastewater services.

## KEY TASKS PERFORMED

During the course of this study, the research team completed many tasks and sub-tasks. Some of the major tasks that were completed by the research team during this study included the following:

- The research team met with the AwwaRF Project Advisory Committee to establish a working relationship with the committee, and to reach agreement about the goals, objectives, plan, and schedule for this project.

- Published literature on environmental leadership was reviewed to identify examples of environmental leadership by drinking water utilities, to begin defining environmental leadership, and to start identifying environmental stakeholders.

- Drinking water utility managers and representatives of environmental stakeholder groups were interviewed to assess their views on the role of a drinking water utility in environmental leadership, the activities that define environmental leadership by a water utility, the barriers that prevent a water utility from engaging in environmental leadership, and other related issues.

- Residential customers of drinking water utilities were interviewed to assess their views on the role of a drinking water utility in environmental leadership, the activities that define environmental leadership by a water utility, and other related issues. Some of the customers who were interviewed were selected from the market served by each of the four utilities that participated in this project. Other customers were selected randomly from around the country.

- An environmental leadership forum was conducted in each of the four participating cities. These forums were attended by drinking water utility managers and representatives of diverse environmental stakeholder groups. The primary objective of these forums was to identify strategies that drinking water utilities could use to become environmental leaders and to begin describing how to use the strategies.

- Preliminary strategies for environmental leadership were developed based on the literature review, interviews with water utility managers and stakeholders, interviews with residential water users, and suggestions made by participants at the environmental leadership forums.

- A second environmental leadership forum was conducted in each of the four cities participating in this project. These forums were attended by the same people who attended the first forums. The objective of the second forum was to obtain feedback about the preliminary guidebook.

- The research team prepared a final report summarizing the results produced by the research team, including a description of how the results were achieved, and, most importantly, resources drinking water utility managers can use to evolve their utilities into environmental leaders.

## CONTENTS OF THE REPORT

The information and strategies presented in this report are intended to provide water utilities with basic guidelines for environmental leadership. The guidebook contains seven chapters and four appendices. The content of each chapter and appendix is briefly described below.

***CHAPTER ONE:*** The first chapter focuses on understanding the concept of environmental leadership. The chapter describes how water utility managers and stakeholders define environmental leadership. The chapter also describes four fundamental characteristics of environmental leadership: environmental stewardship, collaboration, visibility, and the willingness to be first.

***CHAPTER TWO***: An important part of this study involved identifying what organizations outside the water utility industry as well as drinking water utility managers thought water utilities should be doing in the field of environmental leadership. The chapter documents the perceived benefits of environmental leadership, the types of environmental leadership activities that water utilities are currently doing, and why water utilities are not doing more. This chapter also identifies the significant differences in the way water utility managers view their role as environmental leaders compared to the way environmental stakeholders view them.

***CHAPTER THREE***: The third chapter describes the results of a survey that was administered to water utility customers in each of the participating utility markets and to a random sample of residents across the United States. Some of the major findings that are presented in this chapter include: how well informed residents are about environmental issues, how important residents think it is for their water utility to participate in different types of environmental leadership activities, the level of support for various environmental leadership initiatives, and a willingness to pay for environmental leadership initiatives.

***CHAPTER FOUR***: The fourth chapter describes ways that water utilities can engage employees in environmental leadership activities. This chapter contains examples of methods that have been used by water utilities to involve employees in environmental leadership. The chapter also describes ways to implement and evaluate the effectiveness of these strategies.

***CHAPTER FIVE***: The fifth chapter describes ways that water utilities can collaborate with other organizations on environmental leadership issues. This chapter contains examples of methods that have been used by water utilities to collaborate with other organizations to support environmental leadership initiatives. The chapter also describes ways to implement and evaluate the effectiveness of these strategies.

***CHAPTER SIX***: The sixth chapter describes communication strategies that can be used by water utilities to promote awareness of environmental leadership issues. This chapter contains examples of communication strategies that have been used by water utilities to educate customers and raise awareness of environmental initiatives in their community. The chapter also describes ways to implement and evaluate the effectiveness of these strategies.

***CHAPTER SEVEN***: This chapter describes the environmental leadership initiatives that were implemented by the City of Olathe. During a period of less than two years, the percentage of residents who thought the City was an environmental leader increased from 33% to 71%. This chapter explains how the City of Olathe used many of the strategies described in Chapters 3 thru 6 to dramatically change the way the utility is perceived in the community.

***APPENDIX A***: Contains a list of the types of organizations that should be included in environmental leadership activities that are undertaken by water utilities.

***APPENDIX B***: Contains summaries of the environmental leadership forums that were conducted in each of the participating utility markets. The research team did not develop strategies for all of the ideas that were suggested, so these summaries are provided to document the full range of ideas that were suggested.

***APPENDIX C***: Contains guidance for how to plan and conduct an environmental leadership forum, including a step-by-step guide for planning an environmental leadership forum, sample invitation letters, and a sample agenda.

***APPENDIX D***: Contains tables that show the results of all questions that were administered in each of the participating utility markets and to a random sample of residents across the United States. This appendix also contains a copy of the survey instrument and stakeholder interview script that was used. These documents are intended to give water utility managers tools to administer surveys and stakeholder interviews in their own communities.

## INTENDED AUDIENCE FOR THIS WORK

The environmental leadership strategies contain in this report are intended to provide guidance to all water utility managers, especially those in communities where environmental issues are emerging as important topics for the community.

Water utilities in markets where environmental issues are emerging as major concerns in their community may find value in the lessons learned by utilities in Southern California and other locations. The document is intended to be a broad guide that can be used by a wide range of water utilities.

Since all of the participating utilities are public agencies that provide stormwater, wastewater, and water-related services, some of conclusions in this document may not apply to all water utilities.

# CHAPTER 1
# WHAT IS ENVIRONMENTAL LEADERSHIP?

**OVERVIEW**

The term "environmental leadership" has been used loosely to describe a wide range of environmental initiatives. A literature review conducted by the research team for this project found more than fifty different descriptions and/or definitions of "environmental leadership." One of the broadest definitions, developed by the Kansas Environmental Leadership Program, defined environmental leadership as:

> "...an influence relationship between collaborators that intends real change for the mutual benefit of the collaborators and the environment."

The lack of a concise and easy to understand definition of "environmental leadership" for the drinking water utility industry has made it difficult for water utility managers to determine what their utilities should be doing to help protect the environment. To help build a framework for understanding this important concept as it relates to the water utility industry, the research team for this study conducted more than 150 interviews with environmental stakeholders and water utility managers.

**HOW STAKEHOLDERS DEFINE ENVIRONMENTAL LEADERSHIP**

More than one hundred Canadian and American environmental stakeholders were interviewed by members of the research team to understand how people outside the water utility industry view the role of water utilities in environmental leadership. Those contacted included:

- Environmental activists
- Grange leaders
- State water regulators
- Federal water regulators
- Waste water utility managers
- Manufacturers
- Farmers and ranchers
- Civil engineers
- Microbiologists
- Local public health officials
- Consultants who serve the drinking water industry

More than 90% of those interviewed used some or all of the following statements to at least partially define the term, *environmental leadership*:

> - Involving customers and/or other stakeholder groups outside the water utility industry in making decisions that affect them.

- Proactive management of the factors that degrade water quality or quantity.
- Integrated management of the factors that degrade water quality or quantity throughout an entire watershed.

Some of the environmental stakeholders interviewed further defined environmental leadership by a water utility as having a water utility take the initiative to form collaborative partnerships with other organizations in the watershed that have an interest in water issues with the objective of identifying and reducing/eliminating threats to water quality, water quantity, and to the environment at large.

Other defining characteristics of environmental leadership that were mentioned included:

- Integrating the management of drinking water, waste water, storm water, and bio-solids within the watershed.
- Protecting, preserving and remediating (as needed) the air and soil, as well as water supply. Taking an eco-systems view of the physical environment.
- Continuous efforts to improve: operating efficiencies, costs, user satisfaction, technology, the sustainable supply of safe drinking water, and public health.
- Modifying the behavior of the people who live within a watershed so their actions more frequently contribute to conserving, protecting, preserving and improving the air, soil and water within the watershed.

## HOW WATER UTILITY MANAGERS DEFINE ENVIRONMENTAL LEADERSHIP

In addition to external organizations, the research team also interviewed 61 senior water utility managers from 34 different U.S. states and Canada to better understand how water utility managers define the term environmental leaders. Not surprising, it was difficult for most of the water utility managers who were interviewed to give a clear definition of the term. However, many of those interviewed cited several of the same concepts that were used by environmental stakeholders, including:

- Involving customers and other groups in making decisions that affect them.
- Proactive management of factors that degrade water quality or quantity.
- Integrated management of factors that degrade water quality or quantity throughout an entire watershed.
- Anticipating future changes (examples: the environment [especially due to real estate development], regulatory requirements, public health issues, technology) and preparing to respond to them.
- Frequently informing the public about water issues.
- Managing air and soil quality, not just water sources.
- Integrated management of drinking water, storm water, wastewater, and bio-solids.
- Actively collaborating with relevant other organizations (examples: governmental, volunteer, civic, business) to protect, preserve and restore as needed the air, soil, and water within a watershed.

About one third of water utility managers interviewed said that they often use other words or synonyms for the term, environmental leadership. The most frequently cited synonym for "environmental leadership" was "environmental stewardship."

After analyzing the results of the surveys of drinking water utility managers and stakeholders, the research team conducted an environmental leadership forum in each of the four cities participating in this study. Attending these forums were drinking water utility managers and representatives of many stakeholder groups. The main purpose of these forums was to obtain input to help identify strategies that drinking water utilities could use to strengthen their position as an environmental leader. The strategies recommended by forum participants were analyzed by the research team to further identify the defining characteristics of environmental leadership.

## FUNDAMENTAL CHARACTERISTICS OF ENVIRONMENTAL LEADERSHIP

The research team first analyzed and then synthesized the findings from the literature review, interviews with stakeholders and utility managers, and feedback from the utility managers and stakeholders who attended environmental leadership forums. This process identified four defining characteristics of environmental leadership for drinking water utilities. Although each characteristic is independently important, a drinking water utility must simultaneously display all four characteristics to truly be an "environmental leader."

### 1. Environmental Stewardship

Environmental stewardship is the most fundamental characteristic. It implies that a water utility is concerned about the natural environment and is proactively taking steps to protect, preserve, and restore the natural environment and to conserve natural resources.

In each phase of this study (literature review; interviews with utility managers, stakeholders, and residential customers; the forums), one of the common themes (and the most prominent) was that drinking water utilities should operate in ways that are responsible to the natural environment. It is also clear that the drinking water utility industry has in recent years set an increasingly high priority for itself on decreasing any negative impact on the natural environment while increasing its positive impact.

However, just being a good "environmental steward" does not make a water utility an "environmental leader." In order to be an "environmental leader," a water utility has to do more than just protect the natural environment and comply with federal, state, and other regulations. It must display the following characteristics, too.

### 2. Collaboration

The term "leadership" implies that water utilities are not working alone. Water utilities that are "environmental leaders" must work with other organizations to collaboratively plan and manage natural resources. A water utility that is working alone is not a leader. The critical importance of drinking water utilities establishing and maintaining collaborative partnerships with other organizations was another common theme the research team found in each phase of this study. Such partnerships help water utilities acquire additional resources (examples: funds, specialized expertise, supplies, equipment) that are required to achieve an objective. In addition, when these partnerships include organizations whose objectives are different from those of other organizations in the partnership, the decisions that ultimately are agreed to will result in fewer adversarial relations with (potentially highly visible) stakeholder groups.

## 3. Visibility

Almost by definition, 'leadership' requires visibility. Water utilities that are "environmental leaders" will let the public and other organizations know what they are doing and why they are taking steps to conserve, protect, preserve, and restore the natural environment. A water utility that is not educating its customers, employees, and other organizations in its community about what it is doing to help protect, preserve, restore and conserve the natural environment (and what those who live and/or work in the watershed could do as well) is not being an environmental leader. A drinking water utility that engages in environmental stewardship and develops collaborative partnerships, but is *invisible,* has lost two significant opportunities. First, it cannot serve as a model of environmentally friendly and responsible behavior to others because the utility's actions will not be known to people outside the utility. Such a utility will be unable to use one of the most effective methods for changing human behavior, observation of models, to help increase the frequency with which those who live and/or work in its service area act to conserve, protect, preserve, and restore the natural environment. Secondly, a utility that is invisible will be subject to the same kinds of criticisms as the research team found in its interviews with stakeholders and residential customers. In effect, a drinking water utility that engages in environmental stewardship and develops collaborative partnerships, but is *invisible,* will not maximize its benefits from investing in environmental stewardship and collaborative partnership activities because the secondary impact of environmental leaders – that of changing the attitudes and behaviors of customers and others in the utility's service area – will be missed.

## 4. Willingness to Be First

It is difficult to identify an individual or an organization that is considered to be a leader who does not initiate action. All leaders in business, industry, politics, and the military initiate actions. In a national survey that was conducted as part of this study, one of the most frequently mentioned reasons that customers did <u>not</u> think their water utility was an environmental leader was because they thought their utility was just complying with federal or state regulations. If a water utility is going to be perceived as an "environmental leader," its customers and other stakeholders must observe a willingness of the water utility to engage in environmentally responsible actions first – before they are required. While this may entail some risk by a water utility, it is essential if others are to view the utility as a leader in this field. The degree of risk can be reduced by the use of collaborative partnerships to engage customers and relevant other stakeholders from outside the water utility in making decisions about the utility's future initiatives to conserve, protect, preserve and restore the natural environment.

## SUMMARY

Although environmental leadership implies many attributes, four characteristics are essential for water utilities:

- Environmental Stewardship
- Collaboration
- Visibility
- Willingness to be First

Being perceived as an environmental leader means more that just taking care of the environment. In order to be perceived as a leader, water utilities must be willing to take the initiative to work with other organizations in a visible way that allows their customers, community leaders, and special interest groups to see that the utility is truly committed to the protection and preservation of environmental resources.

# CHAPTER 2
# HOW ENVIRONMENTAL STAKEHOLDERS AND WATER UTILITY MANAGERS VIEW THE ROLE OF A WATER UTILITY IN ENVIRONMENTAL LEADERSHIP

## OVERVIEW

An important part of this study involved identifying what organizations outside the water utility industry as well as drinking water utility managers thought water utilities should be doing in the field of environmental leadership. The primary goals of this part of the study were to better understand (1) the perceived benefits of environmental leadership, (2) the types of environmental leadership activities that water utilities were doing and (3) why water utilities were not doing more.

The research team conducted telephone interviews with 61 senior water utility managers from 34 different U.S. states and Canada and with 104 environmental stakeholders representing diverse groups, including:

- Environmental activists
- Grange leaders
- State water regulators
- Federal water regulators
- Waste water utility managers
- Manufacturers
- Farmers and ranchers
- Civil engineers
- Microbiologists
- Local public health officials
- Consultants who serve the drinking water industry

## BENEFITS OF ENVIRONMENTAL LEADERSHIP FOR WATER UTILITIES

### Benefits According to Stakeholders

Almost all of the stakeholders interviewed indicated that water utilities benefit from environmental leadership activities. The benefits identified by stakeholders included:

- Lowering treatment costs by preventing contaminants from entering the water supply. ("An ounce of prevention is worth more than a pound of cure.")
- Being better able to make a long term contribution to public health by reducing the potential for contaminants to enter the water supply.
- Reducing the possibility of adversarial relationships developing between the utility and stakeholder groups.
- Increasing the likelihood that consumers will approve bonds and increased user fees to pay for projects that protect or preserve the environment and to improve the infrastructure.

**Benefits According to Water Utility Managers**

Nearly all of the water utility managers interviewed were able to identify at least one benefit of environmental leadership. The most frequently mentioned benefits included:

- Greater public confidence in the water utility. Bond issues are more likely to be passed because the public believes the money will be spent wisely.
- Adversarial relationships are less likely to occur between the utility and stakeholder groups, which will save time and other resources that are often diverted to respond to "unhappy" groups.
- Greater likelihood of complying with regulatory requirements, now and in the future.
- Improved ability to sustain a supply of safe drinking water that exceeds projected demand far into the future.
- Lower treatment costs because there are fewer pollutants in the water source.
- Less likelihood of fines for non–compliance with regulatory requirements.
- Better relationships with the media; less likely to be subjected to critical scrutiny by the media.

## REASONS DRINKING WATER UTILITIES ARE NOT MORE INVOLVED IN ENVIRONMENTAL LEADERSHIP

**Reasons for Not Being More Involved According to Stakeholders**

Although most of the stakeholders interviewed thought water utilities should be more involved in environmental leadership, many were skeptical about the level of commitment water utilities would actually make to help protect the environment. More than 80% of the stakeholders interviewed identified one or more obstacles that they thought would prevent water utilities from engaging in environmental leadership activities more often than they do. The barriers that were mentioned most often included:

- Insufficient resources: personnel with relevant expertise, budget, time.
- Political interference.
- Frequent turnover among political leaders (due to elections) results in the constant requirement to inform new political leaders; it is difficult to make progress when there is frequent turnover among political leaders.
- Lack of significant state or federal financial incentives to engage in environmental leadership.
- The parochial vantage point of water utility managers who limit their focus to compliance with current regulations, the present time period, and their service territory.

**Reasons for Not Being More Involved According to Water Utility Managers**

Although the majority of the water utility managers interviewed thought environmental leadership was important, not all of the utilities represented in the interviews were actively engaged in environmental leadership activities. Two thirds of the managers said there were barriers and obstacles that prevent their drinking water utilities from engaging in environmental leadership activities more often. The most commonly cited obstacles were limited resources of time, money, and personnel with relevant expertise. Other reasons for **not** being engaged in environmental leadership activities included:

- Their area of responsibility is limited to a service territory, which may not include the entire watershed.
- They are too busy trying to meet current regulatory requirements rather than exceeding them and preparing for more stringent standards in the future.
- Many utility managers view the management of drinking water, storm water, wastewater, and bio-solids as separate entities, rather than managing these forms of water in an integrated, unified manner.
- Lack of personnel with the skills required to establish and maintain collaborative, productive relationships with individuals and organizations whose objectives may be different from those of the water utility.
- Lack of personnel with the skills required to effectively manage performance at the level of organization, work process, and individual employee.
- Senior managers in the utility do not see value in establishing collaborative partnerships with other relevant organizations to protect, preserve and restore the environment within the watershed. Environmental leadership activities are considered to be a waste of time and money.

## OTHER FINDINGS FROM INTERVIEWS WITH DRINKING WATER UTILITY MANAGERS

- Slightly less than half the water utility managers said their organizations had developed specific strategies, goals, and plans for environmental leadership. However, when asked to describe these strategies, goals, and plans, the managers did not provide specific details. Informal questioning of managers showed that only about 10-15% of the water utilities that were represented in the interviews had formal documents about their environmental leadership activities. Strategic plans and mission/vision statements seldom including anything about protecting the environment.

- Most of the managers surveyed could not identify a water utility that they judged to be a good example of environmental leadership.

- Most of the managers thought their water utility should collect information about environmental leadership from the general public. The information obtained by their organizations included:
  - What environmental leadership activities the public expects the utility to do.
  - How well the utility performs the environmental leadership activities the public expects it to do.
  - What characteristics the public expects their drinking water to possess.
  - How well the utility produces water with the characteristics the public expects.
  - Issues and concerns about water quality.

- Although some water utilities were not currently engaged in environmental leadership activities, 85% of water utility managers interviewed said a water utility would experience disadvantages if it does not take at least some action. The two major disadvantages for not participating in environmental leadership activities were:

    1. Higher operating costs in the long run due to higher cost of treating water to remove contaminants that could have been kept out of the water source by using environmental leadership.
    2. Potential for the public and key stakeholder groups to distrust the water utility or develop an adversarial relationship with it. Rate payers could be less likely to vote for bond issues required to fund improvements to the system as a result of this mistrust.

## ENVIRONMENTAL LEADERSHIP ACTIVITIES THAT STAKEHOLDERS THOUGHT WATER UTILITIES SHOULD BE DOING

Stakeholders had many suggestions about the kinds of environmental leadership activities appropriate for a water utility, including:

- Informing the public regularly and frequently about the many water issues that effect them.
- Soliciting diverse stakeholder groups (including those with competing objectives) for ideas to protect and preserve the water, air and soil resources within an entire watershed. And to treat polluted air, soil, and water.
- Integrated management of drinking water, wastewater, storm water, and bio-solids throughout a watershed.
- Assigning a high priority to sustaining a supply of safe drinking water that exceeds the requirements of a growing population for decades into the future.
- Actively participating in organizations that plan and manage land uses within a watershed.
- Actively participating in relevant professional organizations (e.g., AWWA, AwwaRF, etc.).

## TYPES OF ENVIRONMENTAL LEADERSHIP ACTIVITIES THAT WATER UTILITY MANAGERS INDICATED THEY ARE DOING

Most of the utility managers interviewed said their organizations were involved in at least some type of environmental leadership activity or had plans to do so in the coming year. The types of activities that managers indicated their utility was currently doing included:

- Active participation in professional organizations such as AWWA and AwwaRF.
- Providing the public with some information about water issues.
- Actively soliciting comments from the public and stakeholder groups about environmental issues and concerns.
- Collaborating with other organizations to manage sources of drinking water within a watershed.

- Buying land to serve as a buffer around the water source.
- Protecting and preserving water sources.
- Implementing water conservation processes and recycling wastewater.

## COMPARATIVE ANALYSIS OF THE MAIN FINDINGS FROM INTERVIEWS WITH STAKEHOLDERS AND WATER UTILITY MANAGERS

A major finding of the interviews with stakeholders and water utility mangers is that important similarities exist between the two groups regarding the role of a drinking water utility in environmental leadership.

Stakeholders and water utility managers cited similar benefits that a drinking water utility would receive by engaging in environmental leadership. For example, both groups said that news media and stakeholders would be less likely to become adversaries if a utility engaged in environmental leadership. The two groups also agreed that a drinking water utility must overcome many obstacles for it to become an environmental leader. Finally, the environmental leadership activities recommended by stakeholders includes actions that some utility managers indicated their utilities was already taking.

A second key finding of the interviews with stakeholders and water utility managers involved a major difference in the way the two groups perceived the level of environmental leadership by water utilities. The majority of the water utility managers that participated in the interviews indicated that they thought environmental leadership was important, and almost 90% of those interviewed indicated that their utility was engaged in some form of environmental leadership activity (Figure 2.1). However, external organizations were generally not aware of the activities being conducted by water utilities. Only 10% of stakeholders said they were familiar with a water utility's environmental leadership activities (Figure 2.1).

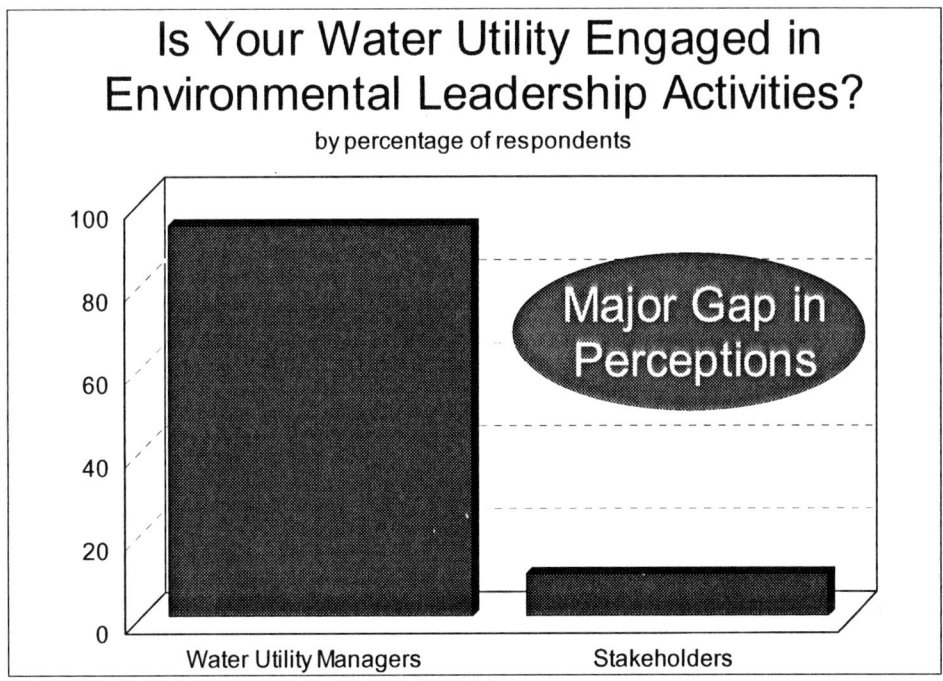

**Figure 2.1: Gaps in Perceptions About Environmental Leadership**

## SUMMARY

The water utility managers and environmental stakeholders who participated in the interviews generally agreed that it was important for water utilities to engage in environmental leadership activtites. Although both groups agreed about the importance of the issue, there was a major gap in the perceived level of environmental leadership being undertaken by water utiltiies. More than nine in ten of the water utility managers interviewed thought their utility was engaged in environmental leadership activities while less than one in ten of the stakeholders interviewed felt that way. This difference defines a critical challenge for the managers of drinking water utilities.

This gap appears to stem from a difference in the way water utility managers perceive their role as environmental leaders compared to the expectations that stakeholder groups have of the industry. Water utility managers typically place a high level of importance on environmental stewardship functions. Although the stewardship function is important, it is only one of four major aspects of environmental leadership as dscribed in Chapter One of this report.

By not openly collaborating with other organizations, visibly educating customers about environmental issues, or being the first organization to take action on selected environmental issues, water utility managers are often viewed as complying with environmental regulation rather than leading efforts to protect the environment.

# CHAPTER 3
# HOW CUSTOMERS PERCEIVE ENVIRONMENTAL LEADERSHIP

## OVERVIEW

During the fall of 2003, the research team designed and administered a survey to a random sample of 400 residents in each of the four participating utility markets and to a random sample of 800 residents from across the United States for a total of 2400 completed surveys. The results for each utility market had a precision of at least +/-5% at the 95% level of confidence. The results of the national survey had a precision of at least +/-3.5% at the 95% level of confidence. The survey was designed to assess perceptions that the general public has about environmental leadership issues in each of the four participating utility markets. The national survey provided a comparative benchmark for evaluating the results from each of the participating utilities. The four utility markets participating in this project included Fort Lauderdale, Kansas City, Olathe and San Diego. Due to geographical proximity, survey data for Kansas City, Missouri, and Olathe, Kansas, were combined and are shown as "Kansas City Metro" in this report.

## GENERAL FINDINGS

### Awareness About Environmental Issues

Although nearly three-fourths (72%) of the respondents from the national survey indicated that they were either "very informed" or "somewhat informed" about environmental issues, only 17% of the respondents indicated that they were "very informed." (Figure 3.1) Respondents who lived in areas where environmental issues have received more attention in the media, such as California and Florida, were significantly more likely to indicate that they were "very informed" than those who lived in places where environmental issues have not been as visible. For example, residents of San Diego were more than twice as likely as residents of the Kansas City metropolitan area to think they were "very informed" about environmental issues (26% San Diego vs. 12% Kansas City).

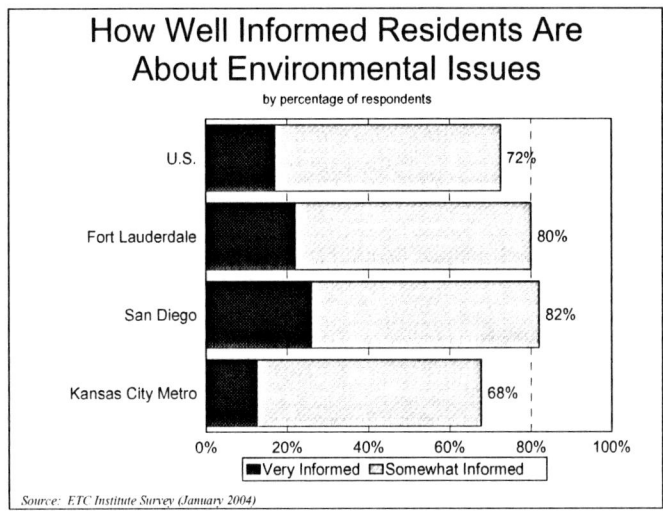

**Figure 3.1: How Well Informed Residents Are About Environmental Issues**

**Importance of Water Utility Being an Environmental Leader**

The level of importance that residents placed on environmental leadership was generally high across the United States. The level of importance that residents placed on environmental leadership was not strongly related to the actual number of environmental problems or challenges that had occurred in a community. Residents living in areas with major environmental challenges were just as likely to think it was important for their water utility to be an environmental leader as residents living in areas with fewer environmental concerns. The level of importance respondents from the national survey placed on a water utility being an environmental leader was 91%, based upon the combined percentage of "very important" and "somewhat important" responses (Figure 3.2). Among participating utilities surveyed, the level of importance ranged from 90% in San Diego to 96% in Fort Lauderdale. Nationally, residents were 21 times more likely to think it was "very important" for their water utility to be an environmental leader than they were to think it was "not important."

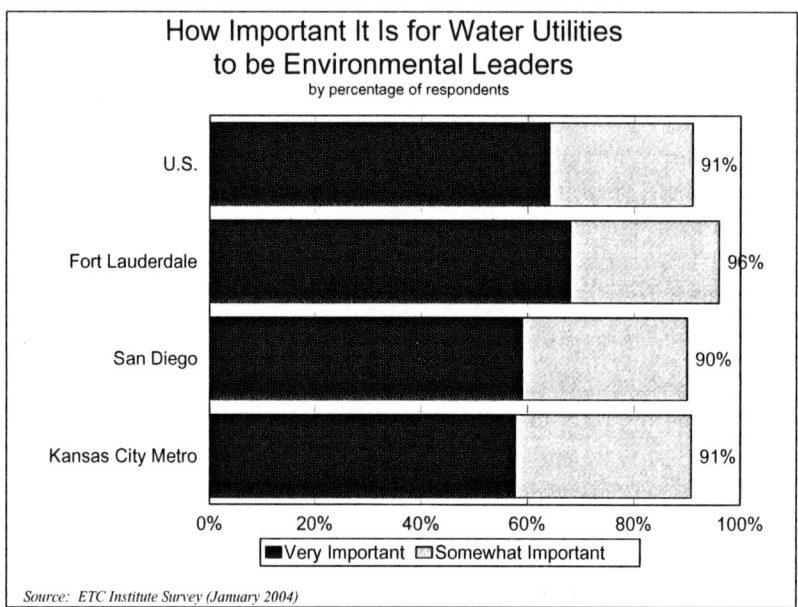

**Figure 3.2: Importance of Water Utilities Being Environmental Leaders**

**Overall Attitude of Communities Toward the Environment**

Eighty-three percent (83%) of respondents from the nation-wide survey thought their community would be willing to adopt new ways to protect the environment, based upon the combined percentage of "very willing" and "somewhat willing" responses. Among participating utilities surveyed, willingness of the community to adopt new ways to protect the environment ranged from 79% in Metro Kansas City to 82% in Fort Lauderdale.

**Is Your Water Utility an Environmental Leader?**

Thirty-one percent (31%) of respondents from the nation-wide survey thought their water utility <u>was</u> an environmental leader. Only 13% of the respondents thought their water utility was not an environmental leader while the majority (56%) responded "Don't know." Findings from the local surveys closely resembled those from the national survey. Among the participating utilities, respondents who thought their water utility was an environmental leader ranged from 30% in San Diego to 33% in Metro Kansas City while those responding "Don't know" ranged from 53% in San Diego to 60% in Metro Kansas City. The research team had originally hypothesized that residents in markets, such as San Diego and Fort Lauderdale, where environmental issues were more visible to the general public would be more likely to have an opinion about their utility than residents in markets where environmental issues were not as visible. The similarity in the results on this issue among participating utilities and the national survey suggest that the expectations that residents have for their water utility may rise as environmental issues become more important to their community. In other words, residents who live in areas like San Diego where environmental issues are more visible may expect more from their water utility than residents in places where environmental issues are not as visible, like Olathe, Kansas. By starting environmental leadership initiatives before environmental issues are a major concern in their community, water utility mangers will find it easier to get residents to view their utility as an environmental leader.

## PERCEIVED IMPORTANCE OF WATER UTILITY FUNCTIONS

### Most Important Functions of the Water Utility

Previous AwwaRF research has indicated that customers think providing safe drinking water is one of the most import functions of a water utility, so the research team was not surprised that respondents from the nation-wide survey rated "providing safe drinking water" as the most important function for water utilities. Eighty-eight percent (88%) of those surveyed selected "providing safe drinking water" as one of the most important functions for a water utility to provide. The research team was surprised that residents thought "protecting the environment" (41%) was generally just as important as "providing good tasting drinking water" (42%) and significantly more important than the following:

- providing inexpensive drinking water (27%)
- providing accurate water bills (26%)
- repairing broken water mains quickly (26%)
- providing adequate water pressure to homes (16%)

Figure 3.3 at the top of the following page shows the relative ranking for the importance that residents placed on eight functions provided by water utilities.

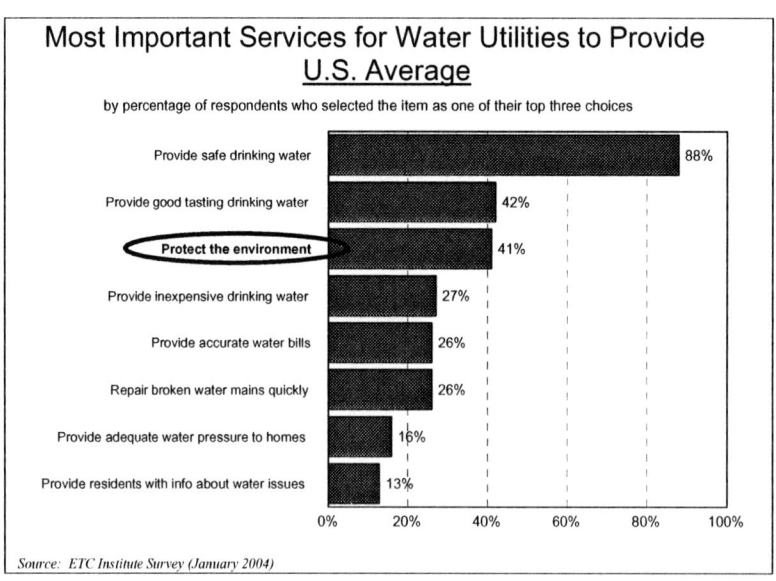

**Figure 3.3: Most Important Services for Water Utilities to Provide**

## IMPORTANCE PLACED ON ENVIRONMENTAL LEADERSHIP ACTIVITIES

**Importance of Cooperating with Other Organizations to Protect Water**

The level of importance respondents from the nation-wide survey placed on cooperating with other organizations to prevent drinking water from being polluted was 95% (for respondents *who had an opinion*), based upon the combined percentage of a 4 or a 5 response on a 5-point scale where 5 was "very important" and 1 was "not important at all". Among participating utilities surveyed, the level of importance ranged from 94% in Metro Kansas City to 98% in Fort Lauderdale.

**Importance of Anticipating Impact of Future Development on Water**

The level of importance respondents from the nation-wide survey placed on anticipating the impact of future development on the supply of safe drinking water was 92% (for respondents *who had an opinion*), based upon the combined percentage of a 4 or a 5 response on a 5-point scale where 5 was "very important" and 1 was "not important at all". Among participating utilities surveyed, the level of importance ranged from 89% in San Diego to 97% in Fort Lauderdale.

**Importance of Encouraging People to Protect Water Sources**

The level of importance respondents from the nation-wide survey placed on encouraging people to help protect sources of drinking water from contamination was 94% (for respondents *who had an opinion*), based upon the combined percentage of a 4 or a 5 response on a 5-point scale where 5 was "very important" and 1 was "not important at all". Among participating utilities surveyed, the level of importance ranged from 89% in San Diego to 97% in Fort Lauderdale.

**Importance of Informing the Public to Protect Water Sources**

The level of importance respondents from the nation-wide survey placed informing the public about ways to protect the sources of drinking water was 90% (for respondents *who had an opinion*), based upon the combined percentage of a 4 or a 5 response on a 5-point scale where 5 was "very important" and 1 was "not important at all". Among participating utilities surveyed, the level of importance ranged from 89% in San Diego to 94% in Fort Lauderdale.

**Importance of Cooperating with Other Organizations to Protect ALL Water**

The level of importance respondents from the nation-wide survey placed on cooperating with other organizations to prevent ALL sources of drinking water from being polluted was 90% (for respondents *who had an opinion*), based upon the combined percentage of a 4 or a 5 response on a 5-point scale where 5 was "very important" and 1 was "not important at all". Among participating utilities surveyed, the level of importance ranged from 86% in San Diego to 92% in Metro Kansas City.

**Importance of Encouraging Users in the Region to Conserve Water**

The level of importance respondents from the nation-wide survey placed on encouraging water users in the region to conserve water was 90% (for respondents *who had an opinion*), based upon the combined percentage of a 4 or a 5 response on a 5-point scale where 5 was "very important" and 1 was "not important at all". Among participating utilities surveyed, the level of importance ranged from 88% in Metro Kansas City to 92% in San Diego.

**Importance of Using Environmentally Friendly Equipment and Technology**

The level of importance respondents from the nation-wide survey placed on investing in environmentally friendly equipment and technology was 85% (for respondents *who had an opinion*), based upon the combined percentage of a 4 or a 5 response on a 5-point scale where 5 was "very important" and 1 was "not important at all". Among participating utilities surveyed, the level of importance ranged from 82% in San Diego to 90% in Fort Lauderdale.

**Importance of Helping Fund Protection of the Environment**

The level of importance respondents from the nation-wide survey placed on helping fund initiatives that help protect the environment in their community was 82% (for respondents *who had an opinion*), based upon the combined percentage of a 4 or a 5 response on a 5-point scale where 5 was "very important" and 1 was "not important at all". Among participating utilities surveyed, the level of importance ranged from 73% in San Diego to 87% in Fort Lauderdale.

**Importance of Acquiring Land Near Lakes to Protect from Contamination**

The level of importance respondents from the nation-wide survey placed on acquiring land near lakes and streams to protect water sources from contamination was 81% (for respondents *who had an opinion*), based upon the combined percentage of a 4 or a 5 response on a 5-point scale where 5 was "very important" and 1 was "not important at all". Among participating utilities surveyed, the level of importance ranged from 77% in San Diego to 88% in Fort Lauderdale.

**Importance of Cooperating with Other Kinds of Environmental Organizations**

The level of importance respondents from the nation-wide survey placed on cooperating with other organizations to protect air quality and other aspects of the environment that are not directly related to water quality was 80% (for respondents *who had an opinion*), based upon the combined percentage of a 4 or a 5 response on a 5-point scale where 5 was "very important" and 1 was "not important at all". Among participating utilities surveyed, the level of importance ranged from 67% in San Diego to 83% in Fort Lauderdale.

**Importance of Managing All Forms of Water in an Integrated Manner**

The level of importance respondents from the nation-wide survey placed on managing all forms of water in an integrated manner rather than separately was 75% (for respondents *who had an opinion*), based upon the combined percentage of a 4 or a 5 response on a 5-point scale where 5 was "very important" and 1 was "not important at all". Among participating utilities surveyed, the level of importance ranged from 73% in San Diego to 82% in Fort Lauderdale.

**Most Important Environmental Leadership Activities**

Respondents from the nation-wide survey *who had an opinion* indicated that the most important environmental leadership activities for water utilities, based upon the combined percentage of a 4 or 5 response on a 5-point scale where 5 was "very important" and 1 was "not important at all", were cooperating with other organizations to protect YOUR water (95%), encouraging people to protect sources of water (94%), and anticipating the impact of future development on water (92%). Respondents from the nation-wide survey placed less importance on managing all forms of water in an integrated manner. (75%).

## OVERALL SATISFACTION WITH ENVIRONMENTAL LEADERSHIP ISSUES

**Satisfaction with Water Utility's Cooperation with Other Organizations**

The level of satisfaction respondents from the nation-wide survey had with how well their water utility cooperated with other organizations to protect and preserve sources of drinking water was 64% (among respondents *who had an opinion*), based upon the combined percentage of "very satisfied" and "satisfied" responses. Among participating utilities surveyed, the level of satisfaction ranged from 60% in San Diego to 68% in Fort Lauderdale.

**Satisfaction with Water Utility's Proactive Protection of Drinking Water**

The level of satisfaction respondents from the nation-wide survey had with how well their water utility proactively protected their drinking water was 55% (among respondents *who had an opinion*), based upon the combined percentage of "very satisfied" and "satisfied" responses. Among participating utilities surveyed, the level of satisfaction ranged from 54% in Metro Kansas City and San Diego to 59% in Fort Lauderdale.

**Satisfaction with Water Utility's Education Efforts**

The level of satisfaction respondents from the nation-wide survey had with how well their water utility educates people about ways to protect and preserve sources of drinking water was 37% (among respondents *who had an opinion*), based upon the combined percentage of "very satisfied" and "satisfied" responses. Among participating utilities surveyed, the level of satisfaction ranged from 37% in Metro Kansas City to 42% in Fort Lauderdale.

**Satisfaction with How Often Water Utility Seeks the Ideas of Residents**

The level of satisfaction respondents from the nation-wide survey had with how often their water utility asks residents for their ideas to protect and preserve sources of drinking water was 32% (among respondents *who had an opinion*), based upon the combined percentage of "very satisfied" and "satisfied" responses. Among participating utilities surveyed, the level of satisfaction ranged from 25% in Fort Lauderdale to 29% in San Diego.

**Overall Satisfaction with Water Supplier**

Overall satisfaction with the water supplier for the nation-wide survey was 66% (among respondents *who had an opinion*), based upon the combined percentage of "very satisfied" and "satisfied" responses. Among participating utilities surveyed, the level of overall satisfaction ranged from 62% in Fort Lauderdale to 69% in Metro Kansas City.

**WILLINGNESS TO PAY**

**Willingness to Pay More If Water Utility Is an Environmental Leader**

Forty-six percent (46%) of respondents from the nation-wide survey indicated that they would be willing to pay a little more if their water utility was recognized as a leader in protecting and preserving sources of drinking water (Figure 3.4). Among the participating utilities surveyed, the percentage of respondents who indicated that they would be willing to pay more ranged from 43% in Metro Kansas City to 53% in San Diego.

**How Much Respondents Would Be Willing to Pay for Environmental Leadership**

Forty-two percent (42%) of the respondents from the nation-wide survey indicated that would be willing to pay up to $2 a month if their water utility was recognized as a leader in protecting and preserving sources of drinking water. Conversely, thirty-three percent (33%) of the respondents from the nation-wide survey were not willing to pay more if their water utility was recognized as a leader in protecting and preserving sources of drinking water. Among participating utilities surveyed, the percent of respondents willing to pay up to $2 a month ranged from 37% in Metro Kansas City to 50% in San Diego while those not willing to pay more ranged from 37% in Fort Lauderdale to 34% in San Diego (see figure on the following page). Although a majority of those surveyed were not willing to pay for environmental leadership, those who thought their water utility was an environmental leader (62%) were significantly more willing to pay for investments to protect the environment than those who did not think their utility was an environmental leader (28%). This suggests that one of the most important benefit to a water utility of being perceived as an environmental leader is that customers will be more willing to pay for environmental initiatives.

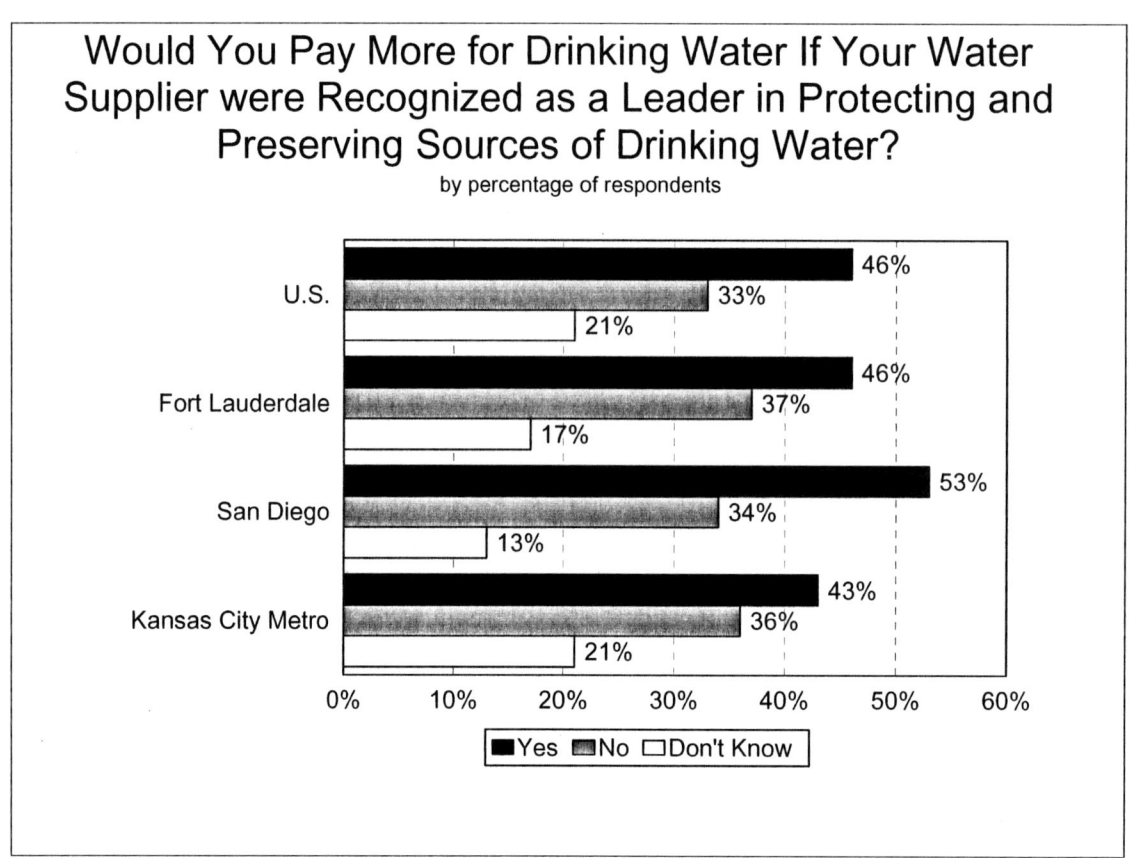

Figure 3.4: Willingness to Pay

## SUPPORT FOR ENVIRONMENTAL LEADERSHIP INITIATIVES

**Support for Various Environmental Leadership Initiatives**

The highest levels of support respondents from the nation-wide survey had for various environmental leadership initiatives, based upon the combined percentage of "very supportive" and "supportive" responses of respondents *who had an opinion,* were investing in water treatment equipment (84%), encouraging local governments to adopt codes for developers (78%), and working with other organizations to restrict development (76%) (Figure 3.5). Respondents were generally less supportive of restricting the times and days that residents can water their lawns or fill swimming pools (56%).

**Support for Investing in Water Treatment Equipment**

The level of support respondents from the nation-wide survey had for investing in water treatment equipment to ensure that the quality of drinking water exceeds federal safety standards was 84% (among respondents *who had an opinion*), based upon the combined percentage of "very supportive" and "supportive" responses. Among participating utilities surveyed, the level of support ranged from 78% in San Diego to 89% in Fort Lauderdale.

**Support for Encouraging Local Governments to Adopt Codes for Developers**

The level of support respondents from the nation-wide survey had for encouraging local governments to adopt codes that require developers to install water saving features to help conserve water use was 77% (among respondents *who had an opinion*), based upon the combined percentage of "very supportive" and "supportive" responses. Among participating utilities surveyed, the level of support ranged from 71% in Metro Kansas City to 79% in Fort Lauderdale.

**Support for Working with Other Organizations to Restrict Development**

The level of support respondents from the nation-wide survey had for working with other governmental organizations to restrict development and recreational activity near lakes and other water ways that are used as sources of drinking water was 76% (among respondents *who had an opinion*), based upon the combined percentage of "very supportive" and "supportive" responses. Among participating utilities surveyed, the level of support ranged from 71% in San Diego to 81% in Fort Lauderdale.

**Support for Spending Money on Water Protection Education**

The level of support respondents from the nation-wide survey had for spending money to educate people about ways to protect water resources from pollution was 74% (among respondents *who had an opinion*), based upon the combined percentage of "very supportive" and "supportive" responses. Among participating utilities surveyed, the level of support ranged from 70% in San Diego to 75% in Fort Lauderdale.

**Support for Spending Money on Environmental Protection Education**

The level of support respondents from the nation-wide survey had for spending money to educate people about ways to help protect the environment was 69% (among respondents *who had an opinion*), based upon the combined percentage of "very supportive" and "supportive" responses. Among participating utilities surveyed, the level of support ranged from 63% in San Diego to 75% in Fort Lauderdale.

**Support for Spending Money on Water Conservation Education**

The level of support respondents from the nation-wide survey had for spending money to educate people about ways to conserve water resources was 65% (among respondents *who had an opinion*), based upon the combined percentage of "very supportive" and "supportive" responses. Among participating utilities surveyed, the level of support ranged from 62% in Metro Kansas City to 66% in Fort Lauderdale and San Diego.

**Support for Restricting Times and Days for Watering Lawns or Filling Pools**

The level of support respondents from the nation-wide survey had for restricting the times and days that residents can water their lawns or fill swimming pools was 56% (among respondents *who had an opinion*), based upon the combined percentage of "very supportive" and "supportive" responses. Among participating utilities surveyed, the level of support ranged from 56% in Metro Kansas City to 60% in San Diego.

**Support for Giving Water Utility Authority to Restrict Development**

The level of support respondents from the nation-wide survey had for giving the water utility authority to restrict development and recreational activity near lakes and other bodies of water that are used as sources of drinking water was 75%, based upon the combined percentage of "very supportive" and "supportive" responses. Among participating utilities surveyed, the level of support ranged from 72% in Metro Kansas City to 76% in Fort Lauderdale.

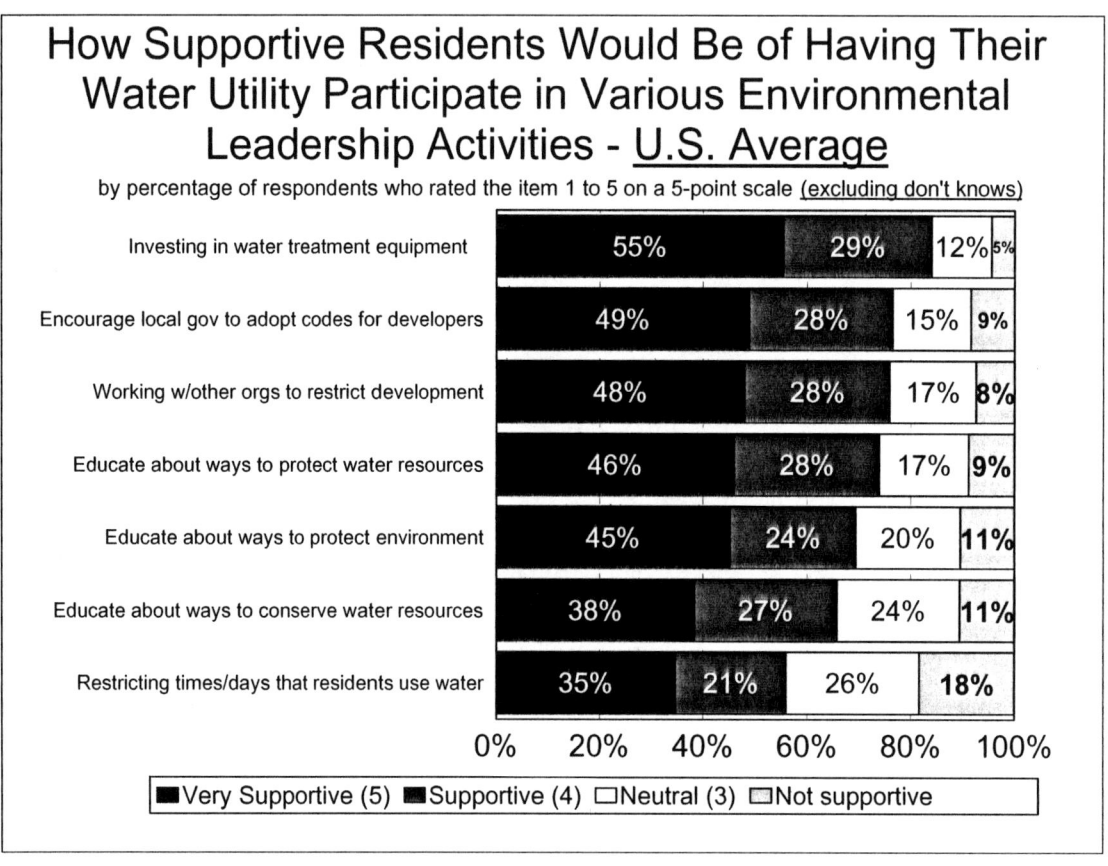

Figure 3.5: Support for Various Environmental Leadership Activities

**SUMMARY**

The results of the residential customer survey indicate that an opportunity exists for drinking water utilities to improve the level of satisfaction residential customers have with their utilities' environmental leadership activities. Examples are described below.

- While 95% of the respondents in the national sample *who had an opinion* said it is very important or important for a drinking water utility to cooperate with other organizations to protect and preserve sources of drinking water, only 64% of these respondents also said they were very satisfied or satisfied with how well their water utility cooperates with other organizations.
- 90% of the respondents in the national sample *who had an opinion* said it is very important or important for a drinking water utility to inform the public about ways to protect drinking water sources. However, only 37% of these respondents said they are very satisfied or satisfied with how well their utility informs its public.
- While 31% of the national sample respondents judge their water utility to be an environmental leader, the vast majority of respondents (69%) said either they did not know whether their utility is an environmental leader (56%) or it is not an environmental leader (13%). Residents in the markets of the four participating utilities had comparable results.

The findings from interviews with residential customers of drinking water utilities were consistent with the main findings of stakeholder interviews reported in Chapter 2. Both groups judged environmental leadership activities to be important for a drinking water utility (Figure 3.6). And both groups generally think that drinking water utilities have an opportunity to improve how well they act as environmental leaders.

In addition, residential customers identified "protecting the environment" as one of the most important services a drinking water utility can provide. Two-fifths (41%) of the respondents in the national sample rated 'protecting the environment' as one of the three most important services a water utility can provide. Only 'providing good tasting drinking water' (42%) and 'providing safe drinking water' (88%) were rated higher.

The following chapters describe specific strategies that can be used by water utilities to become environmental leaders in their communities.

---

## Survey Conclusions

⚬ Environmental Leadership is perceived to be important

⚬ A significant percentage of residents are willing to pay to support it

⚬ Although it is important, most residents don't think of their utility as an environmental leader because they don't know what their utility is doing

⚬ Collaborative efforts to protect water sources from pollution should be a key component of a water utility's long-term strategy

**Figure 3.6: Survey Conclusions**

# CHAPTER 4
# ENGAGING EMPLOYEES TO PROMOTE ENVIRONMENTAL LEADERSHIP

## OVERVIEW

Leaders of organizations in different industries often find it difficult to get their views of their organizations shared by all employees. The greater the number of levels that separate an employee from senior management, the less likely it is that the actions of an employee will reflect the views of senior management. One of the easiest ways that water utilities can become involved in environmental leadership is to engage employees in the process. This chapter presents ways that water utilities can involve employees in environmental leadership.

The primary objective of this strategy is for the job performance of each utility employee as well as the results produced by the utility as a whole to contribute to the protection, preservation, conservation, and restoration of the water sources and of the air and soil that impacts water sources within the watershed or region.

This strategy is recommended for use when a drinking water utility senior manager chooses to improve the utility's positive impact on the environment. The strategy of Engaging Employees gives the entire management team of a drinking water utility a comprehensive, integrated, and systematic process by which to manage the factors that control the results produced by individuals, work processes, and of the utility as a whole. This strategy focuses the impact of these factors on outcomes that illustrate environmental leadership.

## EXAMPLES OF THIS STRATEGY

Water utilities across the United States have used multiple tactics to actively involve their employees in environmental leadership. Below and on the following pages are some of the ways these utilities have encouraged their employees to be more concerned and proactive about protecting the environment.

### Integrating Environmental Leadership into Organizational Mission Statements and Strategic Plans

Examples of this initiative are described below.

- One of the participating utilities in this study developed a mission statement that dedicates the department to being an environmental leader. This utility also created a strategic planning committee that includes a representative from each of the department's six operating units. One of the core functions of the strategic planning committee is to solicit employee suggestions to improve work processes and to ensure that work functions are consistent with the organization's environmental leadership goals.

- Another of the participating utilities in this study created a mission statement that dedicates the Department to provide ". . . Safe, Reliable, Cost-Effective Water and Outstanding Customer Service in an Environmentally Sensitive Manner." This mission statement identifies the two critical outcomes of the utility as an organization (water, customer service) and five general measures of these outcomes (safe, reliable, cost-effective, outstanding, and **environmentally sensitive**).

- Another utility created a environmental policy that dedicates the utility to: (1) continually improving work processes and practices to minimize adverse impact on the rural, urban, and coastal environments; (2) complying with legal and regulatory requirements and to water industry standards, (3) preventing environmental pollution due to department operations, (4) minimizing waste of and negative impact on natural resources, (5) communicating to all stakeholders the actions taken by the utility to protect environmental health and pubic safety, and (6) establishing and maintaining an ISO 14001 environmental management system.

**Using Performance Incentives for Complying with Environmental Regulations**

One utility identified during this study created a system for recording the amount of fines paid during the year for non-compliance with regulatory requirements due to employee mistakes. Funds to pay fines are taken from a line item in the department's budget. Whatever remains in this account at the end of the year is used to pay performance bonuses to employees. This creates a clear, explicit contingency between employee actions and the amount of their annual performance bonus.

**Establishing an Environmental Management Position**

One of the participating utilities in this study created a position responsible for ensuring that the department's plans and daily operations contribute to the utility's goal of being an environmental leader. The primary responsibility of this position is to ensure compliance with both regulatory requirements and with the utility's own standards related to the environment. This position is included on all project management teams. Work crews are required to consult with this position before doing any construction, maintenance, or repair (emergency and routine) work in environmentally sensitive areas.

**Modifying the Application Process for New Positions**

Some of the water utilities that participated in this study ask job applicants to describe examples of their accomplishments related to environmental leadership. These examples, along with other data, are used in the applicant evaluation and selection process. By asking applicants about environmental issues during the evaluation and selection process, these utilities reinforce the importance of environmental issues to future employees.

**Distributing Publications on Environmental Leadership Topics**

To help ensure that the job performance of its employees meets the expectations described in its mission statement, one of the participating utilities published and distributed manuals detailing appropriate actions employees should take in a variety of common situations, including managing hazardous materials, dealing with endangered animal and plant species, planning and managing construction projects, managing storm water run-off, managing risk and safety, and complying with regulatory requirements. These manuals increase the probability that employee job performance will meet the utility's expectations by reducing reliance on fallible human memory.

**Inviting Employees to Public Hearings on Environmental Issues**

Several of the water utilities that participated in the environmental leadership forums that were conducted indicated that they involve employees at all levels and from all functional areas in activities that allow employees to observe the public's reactions to the utility. These activities include attending meetings with the city council, regulators, and the general public as well as staffing booths at events in which these utilities participate or sponsor. By inviting employees to these meetings, employees are more likely to understand the environmental issues that are important to the community and consider the impact that day-to-day decisions may have on these issues.

**Encouraging Energy Conservation and Recycling**

Many drinking water utilities encourage their employees to conserve energy and to recycle at work and at home. For example, some water utilities have placed clearly marked containers at key locations within their facilities in which employees may place items to be recycled. Other utilities use internal publications to communicate the benefits of recycling and conserving energy. These initiatives enable even administrative staff employees to contribute to these utilities living up to the intent of their mission statements.

## HOW TO PREPARE AN ORGANIZATION TO IMPLEMENT THIS STRATEGY

The senior manager of a drinking water utility should initiate and lead this strategy, in collaboration with other top managers. Experience in many industries shows that there usually are some employees at all levels of any organization who will resist using new procedures. To ensure that a drinking water utility gets the full benefit of this strategy, accountability must be established for all members of the management team so that each level of management follows-up frequently and regularly with subordinate managers and supervisors. This will help identify any difficulties that occur during implementation of this strategy and to address them in a timely manner.

The following steps are suggested to prepare a water utility to Engage Employees in environmental leadership.

1. Create a strategy management team to lead the development, implementation, and evaluation of this strategy; include personnel from all levels of the utility's work force (management, non-management, exempt, non-exempt), unions that represent utility employees, and all functional areas of the utility.

2. As appropriate, select one of more of the initiatives that were described on the previous pages in this chapter and create a plan to implement these initiatives in a manner that is specific to your utility.

3. Identify the internal and external resources required to implement your plan.

4. Reach agreement with the body that governs the utility about:
   - The objectives to be accomplished by this strategy.
   - Plan to achieve the strategic objectives.
   - Resources required to achieve the strategic objectives.
   - Any resources from outside the utility that might be required and the probability of obtaining them.

5. Consider forming an Environmental Leadership Advisory Board (or another task force/committee) that will include some persons who are not employed by the water utility to give advice on environmental leadership issues. Organizations from which to recruit people to serve on this Board are listed in Appendix A of this report. Ask new members of the Advisory Board to participate in Steps 6 and 7 below.

6. Update an existing or create a new mission statement that dedicates the utility to protecting the environment. The mission statement should meet these standards:
   - Brief (one to three sentences).
   - Clear and easily understood by employees and the general public.
   - Leads to observable outcomes that can be measured objectively.

7. Determine the critical outcomes, including environmentally friendly outcomes, of the utility as a whole, the measures of these outcomes, and the standards by which these outcomes will be evaluated. These outcomes and their measures should be compatible with the mission statement created in Step 6.

8. Reach agreement with the body that governs the utility about the mission statement, the outcomes the utility as an organization is expected to produce, and the standards by which these outcomes will be evaluated.

9. Use internal publications and meetings to frequently inform the work force about the mission statement; include these topics:
   - What a mission statement is.
   - The reason(s) for a mission statement.
   - The reason(s) for having a mission statement that dedicates the utility to continuously improving its own impact on the environment.
   - The benefits the mission statement created in Step 6 are expected to produce for the utility as an organization, the utility's employees, and for the community.
   - What will be done in the near future to revise internal management practices so the job performance of every employee is compatible with the mission statement.
   - The critical outcomes the utility is expected to produce, the measures of these outcomes, and the standards by which these outcomes will be evaluated.

## HOW TO IMPLEMENT THIS STRATEGY AT THE WORK UNIT LEVEL

The following steps have been developed to help drinking water utility managers implement the strategy of Engaging Employees at the work unit level.

**Step 1:** Within each functional area of utility operations (examples: treatment, distribution, maintenance, etc.), the management team should identify how the function directly or indirectly impacts the environment and the outcomes the function is expected to produce, including environmentally friendly outcomes. Each outcome should meet these standards.

- Be compatible with the outcomes of the utility as an organization that were agreed to in Preparation Step 8.
- Describe a result that is observable to anyone.
- The manager responsible for a function should have primary control over the outcomes of that function, not other managers.
- The outcomes of a function should be the end result of performing all the tasks normally done within that function.

**Step 2:** Determine how to objectively measure each outcome that is identified in Step 1 (examples of measures include accuracy, completeness, rate, volume, timeliness, and labor or material cost).

**Step 3:** Within each function, first line supervisors should:
- Discuss with their subordinate employees how the way these employees perform their daily tasks affects the environment. For employees who deal with water treatment operations or who handle hazardous chemicals, this may be readily apparent. For employees in functions that may have less visible impact on the environment, such as customer service employees, managers should discuss expectations regarding personal behavior at the utility, including efforts to recycle and conserve electricity.

- Inform employees about the outcomes the function is expected to produce and the measures and standards by which performance of the function will be evaluated.

**Step 4:** Annually, review the outcomes, measures, and standards for each function; make appropriate changes, including gradually revising the standards so they are more rigorous. Then, repeat Steps 2-4.

## HOW TO IMPLEMENT THIS STRATEGY AT THE LEVEL OF THE INDIVIDUAL EMPLOYEES

The following steps have been developed to help drinking water utility managers implement the strategy of Engaging Employees at the individual employee level.

**Step 1:** As appropriate, revise the job description of each employee so it describes their tasks and the outcomes for which the individual employee will be held accountable, including environmentally friendly outcomes. Each outcome should meet these standards.

- Be compatible with the outcomes of the function of which the task is a part.
- Describe a result that is observable to anyone.
- The employee responsible for performing a task should have primary control over the outcomes of that task, not other employees.
- The outcomes of a task should be the end result of performing all the steps normally done within that task.

**Step 2:** Determine how to objectively measure each outcome that is identified in Step 1 (examples of measures include accuracy, completeness, rate, volume, timeliness, and labor or material cost).

**Step 3:** Schedule meetings with individual employees to reach agreement about their revised job description.
- Show the job description to the employee and explain the reasons for each outcome, the measures for each outcome, and the standards by which each outcome will be evaluated.
- Ask for and answer the employee's questions.
- Reach agreement about any changes.

**Step 4:** Use internal publications and meetings to frequently inform the work force about:
- What all employees, regardless of their positions, can do to increase their positive impact on the environment, such as recycling at home and work; using public transportation, carpooling, etc.
- What employees in specific positions can do to increase their positive impact on the environment.
- Specific examples of utility employees whose job performance enhanced the environment.

**Step 5:** Annually, review each employee's job description; make appropriate changes, including gradually revising the standards so they are more rigorous. Then, repeat Steps 2 - 5.

## EVALUATING THE IMPLEMENTATION OF THIS STRATEGY AT THE ORGANIZATION LEVEL

Water utilities should follow three basic steps for evaluating the results of using this strategy as described below.

1. Collect performance data frequently and regularly for each measure of the utility's outcomes that were agreed to in Preparation Step 8, then:
   - Compare the actual data against the standard for each measure and determine whether actual performance meets standard, is below standard, or is above standard.
   - Inform employees how well the performance measures for each outcome met their standards.

2. If performance of an outcome meets or exceeds standards, then praise and compliment employees.

3. If the difference between actual performance and a standard shows that an opportunity exists to improve that performance, then reach agreement with relevant employees about the causes of the difference and what will be done to improve performance in the area.

**SUMMARY**

Engaging Employees is a strategy that a drinking water utility can use to focus itself on producing environmentally friendly results. This strategy works by using a systematic process to create an integrated and consistent set of environmentally friendly outcomes for three levels of the utility: employees, work processes, and the organization as a whole.

To implement this strategy, environmentally friendly outcomes that the utility, as an organization, is expected to create are specified. Next, outcomes that are compatible with the organization's outcomes are established for all work processes and functions. Finally, outcomes that are compatible with the outcomes of a work process are established for the individual employees who contribute to that work process. In addition to *expectations*, the strategy also uses these other factors to manage the creation of environmentally friendly outcomes: *feedback* and *recognition*.

By engaging employees in environmental leadership activities, water utilities will be better equipped to meet or exceed the expectations of its diverse stakeholders and the general public about environmental issues.

# CHAPTER 5
# BUILDING PARTNERSHIPS TO PROMOTE ENVIRONMENTAL LEADERSHIP

**OVERVIEW**

Drinking water utility senior managers sometimes find themselves challenged to protect, conserve, and restore the environment and water sources within their service territory or region because their organizations lack adequate resources to do all that is required. This chapter describes methods that will help utility managers form and maintain collaborative partnerships with local and regional organizations for the purpose of jointly managing water sources and related environmental issues within their service area or region.

The primary objective of this strategy is that a drinking water utility, through collaborative partnerships with relevant other organizations, has the resources required to achieve continuous improvement in the quality of air, soil, and water within its service area, watershed, or region.

The strategy of building partnerships helps a drinking water utility acquire through partnerships with other organizations the resources to invest in initiatives that improve the conservation, protection, preservation, and restoration of the environment within the service area, watershed, or region. The resources that become available through these partnerships include cooperation, funds, people, equipment, supplies, and material.

The ways in which the strategy, Building Partnerships, helps a drinking water utility include:

- Partnerships may help generate new and effective ideas to conserve, protect, preserve and restore the environment.

- Initiatives that a utility by itself does not have the resources to accomplish may become possible when the utility has partners.

- Environmental initiatives that are undertaken with partners might result in more positive media coverage than if the utility undertook the same initiatives by itself.

- Efforts to influence the actions of the general public are more likely to be successful since a uniform message (protecting, preserving, conserving, and restoring the environment benefits everyone) is consistently communicated by several key organizations within the region, including the drinking water utility.

The strategy, Building Partnerships, has two parts: (1) Developing Partnerships <u>Inside</u> the Utility's Service Area, and (2) Developing Partnerships With Organizations <u>Outside</u> the Utility's Service Area.

**EXAMPLES OF BUILDING PARTNERSHIPS <u>INSIDE</u> A SERVICE AREA**

**Cooperative Land Use Planning**

A large city water department in Southern California owns two reservoirs that are surrounded by another city. The larger city provides drinking water to residents of the smaller community. Since commercial, residential, and recreational development of land adjacent to these reservoirs could jeopardize the quality of water entering the reservoirs and pose a risk to the water source for both cities, the larger city's water department initiated a partnership with the smaller city to jointly manage land use to protect the reservoirs. One result of the partnership is that representatives of the larger city's water department are members of committees that regulate land use in the smaller community.

**Partnerships with Other Governmental Agencies**

A water utility in upstate New York formed partnerships with local, county, and state agencies in its service area to protect the quality of water in a local lake that was the utility's primary source of water. The lake was located within a 73 square mile watershed that contained parts of three counties and eight communities. This water utility established both formal (codified) and informal agreements with the counties, communities and the New York State Department of Environmental Conservation (NYSDEC). For example, the NYSDEC gave this utility an opportunity to review and comment on applications for permits to disturb the lake shoreline. In another example, representatives from the local water utility were included in County Board of Health reviews of building permits to ensure that proposed development near the lake was compatible with the goal of maintaining high water quality in the lake.

**Partnerships with Schools**

Many utilities have formed partnerships with local schools to create curriculum and lesson plans on environmental and water subject matter for grades K through 12. Three of the participating utilities for this study, including the cities of Olathe (KS), Kansas City (MO), and San Diego (CA), are among water utilities that have worked with local schools.

**Stakeholder Participation in the Development of the CCR**

To meet the EPA's requirement to provide water customers with a Consumer Confidence Report (CCR), a large water utility on the east coast of the United States began using a stakeholder involvement process to design the content and format of its CCR. There were several unanticipated benefits of the stakeholder involvement process. Better communication was established between the water utility and the community it serves. The stakeholder involvement process used with the CCR became a model for public participation in other initiatives. Approximately 100 stakeholders participated in the design of this utility's CCR. The stakeholders included representatives of:

- Public health community
- Consumer groups
- Environmental activists
- Neighborhood leaders drawn from the utility's existing Advisory Neighborhood Commissions
- EPA Region 3
- US Army Corp of Engineers (wholesale supplier of water to the water utility)

# EXAMPLES OF BUILDING PARTNERSHIPS <u>OUTSIDE</u> A SERVICE AREA

## Joint Management of Watersheds

One of the nation's largest water utilities owns reservoirs located in three watersheds with a combined total of 1,970 square miles that are located from 75 to 125 miles outside the utility's primary service area. Decades ago, the water utility established partnerships – and has maintained them - with many different groups in these communities. During the 1990s, agreements with 50 organizations were codified into a formal memorandum of agreement to jointly manage the three watersheds. Distance, alone, has created a challenge to managing these watersheds. More recently, the challenge has intensified due to two factors. First, the Environmental Protection Agency granted the water utility an exception to the requirement to filter surface water. Second, as people have moved into what once was a very rural area, pressure has increased to develop real estate for housing and business and for recreational use of the watersheds.

## Partnerships with Agricultural Interests

A Pennsylvania Water Authority formed a partnership with a nearby watershed management association, individual farmers who live in the watershed, environmental conservation groups, and state government agencies to protect the waters of the water authority's reservoir. The situation is especially challenging as the farmers and other landowners who live within the 140 square mile watershed do not get their water from the reservoir while people who do get water from the reservoir do not live within the watershed. The partnership sponsors and funds numerous educational and outreach programs designed to protect the quality of water entering the reservoir. Some examples are stream bank fencing to prevent livestock from polluting streams while drinking from them, crop rotation, and the use of forested riparian buffers.

## Regional Cooperation for Specific Projects

To create a coordinated, systematic process to manage water supply, flood protection and restore ecosystems in three geographic sub-areas (200 square miles) within a county in southern Florida, one water utility formed two groups to help guide the project. The first group was called the Technical Advisory Committee (TAC). The TAC consisted of organizations and individuals within the utility's service area who had technical expertise relative to creating water balance models. These included consultants and public employees with specialized technical expertise in water supply, waste water, flood control, and ecosystem restoration. The purpose of the TAC was to guide the development of a draft water balance model for each of the three sub-areas that meet a series of technical specifications.

In addition to the TAC, the water utility formed a Project Advisory Committee (PAC) that included representatives from inside and outside the water utility's service area. Some of the types of persons that were included on the PAC were:

- Owners of large tracts of environmentally sensitive land.
- Farmers and other agricultural interest groups
- Utility companies
- Environmental activists

- Representatives of local governments
- Representatives of state government agencies
- Representatives of federal government agencies

The 50 member PAC was divided into three workgroups, one for each of the three sub-areas within the county. The workgroups were charged with developing alternatives for each of the following elements of a water management process:

- Moving water into a sub-area
- Moving water within a sub-area
- Mechanisms to store water within a sub-area
- Moving water out of a sub-area

Unlike the TAC that focused on technical issues, the PAC focused its efforts on policy issues to ensure that recommendations would be politically acceptable to the community. By creating two groups with specific functions and integrating the efforts of each group during the project, the county was able to develop a plan that integrated the views of diverse stakeholder groups with technical expertise and political influence, which gave the recommendations significantly more credibility in the community.

**Sustaining Adequate Supplies of Drinking Water**

Faced with the challenges of sustaining an adequate supply of quality drinking water, this highly urbanized county developed a watershed management plan for a segment of the county north of Atlanta. Extensive public involvement and partnering with external organizations was a key part of the development of the plan. Highlights of the effort are described below.

- Informal working partnerships with the local business association and a community environmental organization were formed early in the study. A primary objective of this partnership was to obtain business compliance with all county codes and regulations to protect water quality. Businesses that met these requirements were recognized by feature reports in local news media and in advertisements. By working with the local business association, the utility was able to involve businesses that were located both inside and outside its service area.

- In addition to traditional public meetings scheduled at a time and place convenient to county officials, presentations were also made at such non-traditional venues as homeowner association meetings and civic club meetings.

- An interactive software program called Lorelei was used at meetings with residents to compare costs, effectiveness and benefits of various optional methods to manage wastewater and to protect source water. The software enabled residents to select different water management methods and highlight an area within the watershed to apply the method. The software would then display the cost of the method and its benefits.

- Multiple stakeholder groups were involved from fifteen communities within the county, including homeowners, real estate developers, farmers, Chamber of Commerce leaders, city and county planners, and environmental activists.

- Stakeholders were involved from the beginning.

## HOW TO PREPARE AN ORGANIZATION TO IMPLEMENT THIS STRATEGY

This strategy is recommended for use when a drinking water utility senior manager wants to improve management of water resources and related environmental issues within a service area or region, but the organization, by itself, lacks the resources to do so. The following steps are suggested to prepare your organization for this strategy.

**Step 1:** Create a strategy management team to lead the development, implementation, and evaluation of this strategy; include personnel from:
- All levels of the utility's work force (management, non-management, exempt, non-exempt).
- Unions that represent utility employees.
- All functional areas of the utility.

**Step 2**: As appropriate, adopt some of the examples provided in this chapter to create a plan that will work for your utility.

**Step 3:** Reach agreement with the body that governs the utility about:
- The objectives to be achieved by this strategy (what should be accomplished).
- Plan to achieve the strategic objectives.

**Step 4:** Form an Environmental Leadership Advisory Board (or other group) to guide the process. Organizations from which to recruit people to serve on this Board are listed in Appendix A of this report. Consult with the Advisory Board during the planning and implementation of this strategy.

## HOW TO IMPLEMENT THIS STRATEGY

Described below are the basic steps to build partnerships <u>inside</u> and <u>outside</u> the utility's service area to conserve, protect, and preserve water sources and to manage related environmental issues.

1. Annually, establish objectives to be achieved within a particular time period by using this strategy.

2. Identify actual current threats and future potential threats to water sources within the service area, watershed, or region; determine all the causes of these threats.

3. Identify where opportunities exist to improve multi-agency and multi-jurisdictional coordinated management of water resources within the service area, watershed, or region and all the causes of these opportunities.

4. Identify **both** the positive and negative economic and other consequences to the utility's service area, watershed, or region, to the utility as an organization, and to the general public of:
   - Conserving, protecting, preserving, and restoring air, soil, and water resources within the service area, watershed, or region.
   - Ignoring actual or potential threats to water sources.
   - Current status of managing environmental issues in the service area, watershed, or region.
   - An improved, joint management of environmental issues in the service area, watershed, or region.

5. Using the data from Steps 2 - 4 above, proactively recruit key stakeholder and relevant other groups and organizations from inside and outside the utility's service area, watershed, or region, to form an ongoing collaborative partnership with the drinking water utility for the purpose of:
   - Protecting all sources of drinking water within the service area, watershed, or region.
   - Jointly identifying and managing environmental and water issues within the service area, watershed, or region from micro-level (example: treatment plant) to macro-level (example: eco-systems management throughout a watershed or region).
   - Increasing community participation in environmentally friendly activities.

6. Identify opportunities to increase the general public's participation in activities that conserve, protect, preserve, and restore air, soil, and water resources. Examples include recycling, stream and field cleanup, watering of landscape, environmental leadership forums.

7. In collaboration with the Environmental Leadership Advisory Board, environmental groups, and relevant other organizations that are concerned about water issues within the service area, watershed, or region, develop a unifying 'environmental' theme and a visual logo to represent the importance of conserving, protecting, preserving, and restoring the environment and water resources. Organize the activities planned in Step 8 within this theme and use the logo when communicating about these activities.

8. Develop, implement, and evaluate action plans to address the threats identified in Step 2 and the improvement opportunities found in Step 3, and to achieve the positive consequences identified in Step 4; include in these plans:
   - Multiple methods that will be used to frequently inform stakeholders and the general public about: the threats to water sources (Step 2), the improvement opportunities (Step 3), what will be done to address the threats and opportunities, and the benefits of these investments (Step 4).

- What will be done to increase public participation in environmentally friendly activities; where possible, design competition into these activities; sponsor and present awards to winners of the competitions; distribute news releases in advance of the activities; inform customers of these events using bill stuffers or other media; invite news media to award presentation ceremonies.

## EVALUATING THE IMPLEMENTATION OF THIS STRATEGY AT THE ORGANIZATION LEVEL

Water utilities should follow three basic steps for evaluating the results of using this strategy as described below.

1. Collect performance data frequently and regularly for each measure of the utility's outcomes that were agreed to in Preparation Step 8, then:
   A. Compare the actual data against the standard for each measure and determine whether actual performance meets standard, is below standard, or is above standard.
   B. Inform employees how well the performance measures for each outcome met their standards.

2. If performance of an outcome meets or exceeds standards, then praise and compliment employees.

3. If the difference between actual performance and a standard shows that an opportunity exists to improve that performance, then reach agreement with relevant employees about the causes of the difference and what will be done to improve performance in the area.

## SUMMARY

Building Partnerships is a strategy that enables a drinking water utility to create and maintain collaborative partnerships with relevant organizations located within and outside the utility's service area for the purpose of pooling resources to jointly conserve, protect, preserve, and restore the environment and to manage related environmental issues. The resources that are made available through these partnerships include inter-agency cooperation, funds, people, supplies, material, and equipment.

These partnerships help a utility to improve environmental quality by mitigating past damage to the environment. The partnerships also help a utility to meet regulatory requirements and to increase public participation in environmentally friendly activities.

# CHAPTER 6
# USING COMMUNICATION TO CHANGE CUSTOMER PERCEPTIONS ABOUT A WATER UTILITY'S ROLE IN ENVIRONMENTAL LEADERSHIP

## OVERVIEW

Drinking water utilities often *react* to situations created by the media, special interest groups, or the general public instead of *proactively* influencing the public discussion of water and environmental issues. The strategy of Changing Customer Perceptions gives managers of drinking water utilities ideas they can use to manage the information about water and environmental issues that is available to the public and key stakeholders.

The objectives of this strategy were developed in response to the following observations that were identified during the course of this study.

- The perceptions of people who live and work within a water utility's service area can be shaped by information they receive about the water utility.

- Perceptions about a water utility's role as an environmental leader are shaped by the knowledge customers have about:
    - Environmental leadership activities of their drinking water utility.
    - Familiarity with water utility operations.
    - Understanding of regulatory requirements.
    - How well they think their drinking water utility complies with regulatory requirements.
    - Awareness of local and regional water issues.

- The majority of customers assign a relatively high priority to conserving, protecting, preserving, and restoring the environment.

The strategy described in this chapter enables drinking water utility managers to improve communication with their customers, stakeholders, and the general public. Some of the reasons for implementing the ideas that are described in this chapter include the following.

- Environmental groups are less likely to criticize the utility because environmentalists participate in utility sponsored activities and/or recognize the utility's role in environmentally friendly projects.

- Environmentally friendly activities sponsored by the utility that are highly visible to the general public with or without media coverage will help to counter-balance media stories that are critical of the utility.

- It increases community participation in activities that conserve water, protect water sources, and that preserve and restore the environment.

The risks of not using the strategy, Changing Customer Perceptions, include the following.

- Media coverage is more likely to be negative than positive.
- Bond issues will be more likely to be rejected by voters.
- Severely critical comments are more likely to be made by customers and other

stakeholders at public meetings.
- Adversarial relations are more likely to exist between the utility and its customers and stakeholders.

## EXAMPLES OF THIS STRATEGY

### On-Going Communications

To proactively manage public evaluation, an Oregon utility developed an ongoing process of communicating with its public. The process has multiple elements.
- Customer relations: reacting quickly and timely to customer complaints or requests for information about water quality.
- When requested, the utility will provide personnel to analyze home water samples.
- The utility provides bill inserts each month on a variety of water related issues.
- Informational brochures about specific projects being undertaken by the water utility are distributed to particular stakeholder and interest groups.
- Monthly water quality newsletter mailed to a select group of more than 200 individuals and organizations. Topics covered include current legislative and regulatory issues and various water supply issues.
- The utility provides speakers to discuss water issues with community and interest groups.
- Public information booths are set up at outdoor fairs held during the summer.
- Using focus groups, stakeholder interviews, and telephone surveys, the utility proactively solicits public comments about water issues.
- The utility established a Water Quality Advisory Committee. Members of this committee represent a variety of community interests related to water. The interested citizens who are members of this committee meet monthly at public meetings to discuss water-related issues and provide advice to the utility.
- The utility maintains close contact with state and federal water regulatory agencies, local public health officials, and other water utilities in Oregon and the Northwest United States.
- The water utility maintains on-going contact with local news media, providing them with tours of the utility's facilities, giving in-depth background information, and responding to requests for information as fully as possible and in a timely manner.
- When potential problems are discovered, information about them is distributed quickly to the public.

### Communicating Directly with Customers

One method to communicate to customers is by demonstrating actions or providing information directly to customers that show the water utility is committed to the environment. Specific examples of these direct actions are provided below.

- **Investments in Environmentally Friendly Facilities**. The City of Olathe (KS) constructed an environmentally friendly building for its Municipal Services

Department. The building and its landscape communicate to citizens – without the use of traditional media - examples of what they can do to conserve water, restore soil resources, recycle materials, and lower their energy costs.

- **Irrigation System Inspections**. A water utility in southern Florida formed a partnership to offer a free irrigation system analysis and evaluation service to residents. At the request of a property owner or manager, this mobile service goes to a property to inspect the irrigation system, then suggest changes that would reduce water usage.
- **Helping Customers Conserve Water and Save Money.** The Water Department of the City of San Diego created software that allows citizens to reduce the amount of water used to irrigate their lawns and shrubs. After accessing this software via the internet, a homeowner enters data about their lawn, trees, shrubs, flowers, irrigation system, and zip code. The software suggests how to irrigate the property with minimal use of water.
- **Brochures on Landscaping and Irrigation**. A water utility in Utah created a free brochure to help homeowners conserve water while irrigating their properties. The brochure contains suggestions for water-wise landscaping, water-saving tips for lawns, and other ideas to reduce the use of water for irrigating a landscape.
- **Website Tools**. A large water utility in Florida uses its website to communicate with its customers. The website contains easily accessed information about this agency, including office locations, operations, management of storm water run-off, tips for property owners to reduce water use, master water plan, annual report, board of directors, meetings of the board of directors, news releases prepared by the agency since 2000, links to relevant other web sites, and the agency's many efforts to conserve, preserve, and restore the natural environment.

**Communicating Through Community Organizations**

Another method involves working with organizations in the community to inform customers about environmental issues. Specific examples of these actions are provided below.

- **Publications in Home Owner Association Newsletters.** The Water Department of the City of San Diego reached agreement with home owner associations and condominium associations that the associations would print in their regular publications articles on water and environmental issues written by the Department.
- **Poster Contests**. In 2001 the City of San Diego Water Department began conducting an annual poster contest for all sixth grade students within the city. The primary objective of the contest is to promote public awareness of the importance of conserving water. Although the name of the contest is changed each year, the common theme is the importance of conserving water resources and reducing water use. Nearly 3,700 students submitted a poster in 2004. Water Department employees judge the submissions and select the winners. To encourage participation, every student who submits a poster is given a certificate of participation. Winners of the contest are given a U.S. Savings Bond, tickets to a science center, a certificate of excellence, and a free meal at a well-known restaurant. To further recognize the winners, but more importantly to publicize the importance of conserving water, the

utility partners with many local organizations in the community to promote the posters. Winning posters are: made available to the public as screen savers for PCs and as calendars; displayed in local art galleries, on billboards, in city facilities, in the lobbies of businesses, and at city and county fairs; and broadcast on the city owned television channel.

**Combining Communicating Through News Media, Direct Communication With Customers, and Communicating Through Community Organizations**

A large utility in Canada used a variety of methods to solicit public involvement in the creation of the utility's Strategic Plan for Water Management. The methods included:

- Public workshops to introduce the process for creating the strategic plan to the general public, to solicit participants' suggestions for issues the plan should cover, and to solicit comments about water management issues the water utility had previously identified.
- Focus groups with key stakeholders, including representatives of agricultural, commercial, industrial, environmentalist, scientific, and residential consumer groups.
- Articles and advertisements published in a local newspaper.
- Soliciting key stakeholders' comments on drafts of the plan.

By using several methods of communication and public input the utility was able to involve more people and build a wider base of support for implementation of the utility's strategic plan after it was adopted.

## HOW TO PREPARE TO IMPLEMENT THIS STRATEGY

This strategy is recommended for use when a drinking water utility senior manager decides to use communication to proactively influence the public's attitudes about the water utility and related environmental issues  The steps for preparing to implement this strategy are described below. Some of the basic steps for initiating this process are similar to the steps that were described in previous chapters.

**Step 1:** Create a strategy management team to lead the development, implementation, and evaluation of this strategy; include personnel from:
- All levels of the utility's work force (management, non-management, exempt, non-exempt).
- Unions that represent utility employees.
- All functional areas of the utility.

**Step 2**: As appropriate, revise the general plan contained in this chapter to create a plan to implement this strategy that is detailed and specific to your utility.

**Step 3:** Reach agreement with the body that governs the utility about:
- The objectives to be achieved by this strategy (what should be accomplished).
- Plan to achieve the strategic objectives.

**Step 4:** Form an Environmental Leadership Advisory Board to give advice on how to successfully communicate information to utility customers and the general public. Organizations from which to recruit people to serve on this Board are listed in Appendix A of this report. Consult with the Advisory Board during the planning and implementation of this strategy.

**Step 5:** Identify key stakeholders within and outside the utility's service area, including: consumer groups (business/commercial; industrial; agricultural; residential), elected officials (city/town, county, state), and regulators (city, county, state, federal). For each stakeholder group, determine:
- Which communication media (e.g., print, broadcast, email, website) are the most effective.
- Which communication methods (e.g., words, pictures) and language are the most effective.

**Step 6:** Determine where opportunities exist to improve what's currently being done to inform the critical stakeholder groups identified in Step 5 and identify all the causes of these opportunities to improve.

**Step 7:** (optional) Within the same government organization (example: city), identify those departments which consistently are the most successful at informing their critical stakeholder groups and the general public; determine the communication practices these departments use.

**Step 8:** Identify those public and/or private drinking water utilities that consistently are successful at informing the critical stakeholder groups identified in Step 5 and the general public; determine the communication practices these utilities use.

**Step 9.** For each critical stakeholder group identified in Step 5, determine what information they expect to get from their drinking water utility about:
- Utility operations (including current performance and future changes).
- Regulatory requirements (current and future).
- How well their drinking water utility complies with current regulatory requirements.
- Local and regional water issues (current and future).
- Environmental leadership activities of their drinking water utility (current, planned, and proposed).
- Storm water issues.
- Waste water issues.
- Sewage issues.
- Other topics related to water and the environment.

**Step 10:** For each critical stakeholder group identified in Step 5, create informational objectives (outcomes) related to:
- Water utility operations (including current performance and future changes).
- Regulatory requirements (current and future).

- How well the water utility complies with current regulatory requirements.
- Local and regional water issues (current and future).
- Environmental leadership activities of the water utility (current, planned, and proposed).
- Storm water issues.
- Waste water issues.
- Sewage issues.
- Other topics related to water and the environment.

**Step 11.** Based on Steps 5 – 10 above:
- Revise existing communication practices as appropriate.
- Modify and adapt as needed the most effective communications practices used by other organizations.
- Develop new communication practices as appropriate.

**Step 12**: For each critical stakeholder group identified in Step 5, determine how to combine communicating through the news media, communicating directly with customers, and communicating through community organizations to most effectively provide the information the group expects to get from its drinking water utility (from Step 9) and achieve the informational objectives described in Step 10.

**Step 13**: Develop, implement, evaluate, modify, and maintain a proactive public relations campaign to:
- Create and sustain public support for environmental leadership activities by the utility.
- Gain the active participation of the public in environmental leadership activities.
- Inform the public about the operations of the drinking water utility.
- Inform the public about the importance of environmental leadership activities by the utility.

**Step 14.** Assign a knowledgeable, articulate, and credible person to serve as spokesperson to communicate the utility's message about environmental leadership to its own employees and to the general public. Select someone from the utility's current work force, hire a new person, or engage a consultant.

## HOW TO IMPLEMENT THIS STRATEGY

The strategy, Changing Customer Perceptions, uses three methods to influence utility customers' knowledge about water and environmental issues: (1) communicating through the news media, (2) communicating directly with customers, and (3) communicating through community organizations. The basic steps to implement this strategy using these three methods are described below and on the following pages.

## Steps to Communicate Through the News Media

**Step 1:** For news media judged effective for the utility's critical stakeholder groups (Preparation Step 5), identify reporters whose 'beat' includes the environment; initiate contact with these reporters with the objective of establishing a personal relation with them.

**Step 2:** Find out what kinds of events interest these reporters and their audiences; determine what information these reporters require in news releases.

**Step 3:** Establish objectives to be achieved by implementing the strategy of communicating through the local news media. Examples might include:
- Number of editorials critical of the utility.
- Number of 'exposes' or critical stories about the utility.
- Number of positive stories about the utility.
- Number of letters to the editor that are critical of the utility.
- Number of letters to the editor that are neutral or positive about the utility.

**Step 4:** When the utility has information relevant to Preparation Steps 9 and 10 to communicate to its customers, stakeholders, and the general public, prepare and distribute news releases that meet the reporters' requirements.

**Step 5:** Invite the reporters identified in Step 1 to attend environmentally friendly public involvement activities sponsored by the utility and community groups and to other relevant events.

## Steps To Communicate Directly With Customers

**Step 1**: Given the stakeholder groups, communication media, and communication methods identified in Preparation Step 5, the informational objectives established in Preparation Step 10, and the communication media controlled by the utility (e.g., printed materials mailed to customers, website, local government television channel, etc.), create a plan to directly communicate with each stakeholder group.

**Step 2:** Design and create the communications required by the plans.

**Step 3:** Implement and evaluate the plans.

**Step 4**: Revise the plans as appropriate.

**Step 5**: Implement the revised plans.

## Steps To Communicate Through Community Organizations

**Step 1:** Identify environmentally friendly public involvement activities the utility <u>currently</u> sponsors or co-sponsors. Examples might include water festivals, poster or essay competitions for students, stream clean-ups, and recycling programs.

**Step 2:** Identify additional environmentally friendly public involvement activities the utility <u>could</u> sponsor in the future.

**Step 3:** Establish objectives to be achieved by implementing the strategy of communicating through community organizations. Examples might include:
- Number of partnerships formed and actively operating by a deadline.
- Number of events co-sponsored with community organizations by a deadline.
- Number of people who attend environmentally friendly events that are co-sponsored by the utility and community organizations.

**Step 4:** Identify community organizations within the utility's service area, watershed, or region that are credible to the majority of the general public; solicit their co-sponsorship, joint planning, and active participation in the activities identified in Steps 1 and 2. Examples include chambers of commerce, schools, environmental activists, and service clubs.

**Step 5:** Where possible, design competition into the activities identified in Steps 1 and 2; actively recruit community organizations to join the utility in sponsoring and presenting awards to winners of the competitions; distribute news releases in advance of the activities; inform customers of these events using bill stuffers or other media; invite news media to the award presentation ceremony.

**Step 6:** In collaboration with the Environmental Leadership Advisory Board and community organizations that are concerned about water and related environmental issues, develop a unifying 'environmental' theme and a visual logo to represent the importance of conserving, protecting, preserving, and restoring the environment and water resources. Organize the activities identified in Steps 1 and 2 within this theme and use the logo when communicating about these activities.

**Step 7:** Collaborate with community organizations that are concerned about water issues to create and sustain implementation of a multi-media, multi-method effort that regularly and frequently communicates the environmental theme to those who live and work within the utility's service area, watershed, or region.

**Step 8:** Determine if any environmental groups (local, regional, national) sponsor awards to drinking water utilities for environmentally friendly accomplishments; determine the qualifications for such awards; manage the utility so it meets the qualifications; apply for awards.

## HOW TO EVALUATE THE IMPLEMENTATION OF THIS STRATEGY

1. For each critical stakeholder group identified in Preparation Step 5, evaluate how well the informational objectives described in Preparation Step 10 have been achieved.

2. If an opportunity exists to improve how well any informational objective(s) have been achieved for any critical stakeholder group:
    - Identify all the real causes.

- As appropriate, revise communication practices (content, media).
- Implement revised communication practices.
- Evaluate revised communication practices.

3. Annually, compare actual data to the objectives; take appropriate corrective action to meet or exceed objectives during the next time period.

4. Establish baseline data for events currently sponsored by the utility; compare future data to baseline; when the data do not show improvement, determine the reasons, then modify the events and/or communications about the events.

5. For each kind of event established in the future, collect baseline data the first time the event occurs; compare baseline data to data from subsequent events; when the data do not show improvement, determine the reasons, then modify the events and/or communications about the events.

## SUMMARY

The strategy of Changing Customer Perceptions helps a drinking water utility to proactively influence the public discussion of environmental and water issues. This strategy works by giving utility managers tools to manage the information about environmental and water issues that is available to the public. This strategy has three parts; communicating through the news media, communicating directly with utility customers, and communicating through community organizations.

By using this strategy, utilities are more likely to increase community participation in activities that conserve, protect, preserve, and restore the natural environment. This strategy also helps to reduce the potential for adversarial relationships between a utility and its stakeholders or news media.

# CHAPTER 7
# BUILDING SUPPORT FOR ENVIRONMENTAL LEADERSHIP INITIATIVES

## OVERVIEW

The research team for this project spent endless hours trying to identify a single strategy for building support for increased funding for environmental leadership activities. The search for this strategy revealed that the key to building support for increased funding is a function of the application of the items in the previous three chapters: (1) engaging employees, (2) taking the initiative to collaborate with other organizations, and (3) proactively communicating information about environmental issues to customers, community leaders, and environmental groups.

One of the four utilities that participated in this study, the City of Olathe, agreed to go beyond just participating in the project. The City of Olathe is a rapidly growing suburb in the southwest portion of the Kansas City area. During the past fifteen years, the City of Olathe has grown from a population of less than 50,000 to more than 120,000 residents. The rapid growth has placed enormous pressure on the local water system.

Prior to the start of this project, the City of Olathe had made a commitment to environmental leadership. As a result, during the course of this project the City implemented many of the concepts and strategies that are presented in this report. This chapter will describe the specific initiatives that were implemented by the City along with the results that were achieved.

## ENGAGING EMPLOYEES TO PROMOTE ENVIRONMENTAL LEADERSHIP

The formal process for engaging employees in environmental leadership was initiated by the former Director of Olathe Municipal Services, William Ramsey, in 1999. He felt that the organization needed a vision that included environmental leadership as a key component of the utility's strategic plan. In order to engage employees in the process, the water utility took three major actions that are briefly described below.

First, senior management administered a survey to employees to gather feedback about the direction of the organization. The results of the survey were used by senior managers to set internal priorities for the organization to respond to employee concerns and build internal support for the utility's strategic goals. The survey is now repeated every two years.

Second, using feedback from the survey, senior managers updated the organization's strategic plan. As part of this update, senior managers made a visible commitment to environmental initiatives, including a stated desire to have the organization become an environmental leader in the community (Figure 7.1). The visible commitment of senior managers was a critical step in creating a culture where employees at all levels placed a priority on protecting the environment.

> **Engaging Employees**
>
> An integral part of the City of Olathe's Municipal Services Department's mission statement is to promote the conservation of Olathe's environment and natural resources. Assuming a leadership role in the community to model environmental stewardship is viewed as one of the primary responsibilities for employees in the department.

**Figure 7.1: Engaging Employees**

Third, the utility created cross-functional teams to review and guide the implementation of the organization's strategic plan. A key purpose of these teams was to keep environmental issues visible to employees. Some of the initiatives that were implemented internally included employee recycling programs and the posting of environmentally friendly messages in work areas. By having these teams meet regularly, senior managers have had many opportunities to openly emphasize the utility's commitment to the environment and train employees how to contribute to the utility's environmental leadership goals.

**BUILDING PARTNERSHIPS**

In the spring of 2003, the City of Olathe launched a major initiative to establish a watershed protection area around one of the City's major sources of water, Lake Olathe (Figure 7.2). For nearly 50 years, Lake Olathe has provided drinking water to the citizens of Olathe. Located in an increasingly urbanizing area, the Lake Olathe Watershed (the 10,640-acre land area that drains to Lake Olathe) has experienced rapid changes in land use. Concerns of increasing sedimentation and nutrient levels negatively impacting water quality in Lake Olathe prompted the City to initiate efforts to protect the watershed.

Working in cooperation with the U.S. Geological Survey (USGS) and the Kansas Department of Health and Environment (KDHE), the City initiated an effort that led to the development of the Lake Olathe Watershed Restoration and Protection Plan. The overall goal of the Lake Olathe Watershed Protection Plan was to protect, enhance, and restore the water quality of the city's two lakes and streams within the 10,640-acre drainage area. Specific watershed restoration and protection goals include:

- preserving Lake Olathe for use as a drinking water supply;
- protecting the Lake Olathe watershed recreational/park areas;
- preserving the wildlife habitat/riparian corridor along the stream system;
- designing effective stormwater management in the watershed;
- preserving the aquatic system as a valuable ecosystem with educational/research benefits; and
- maintaining and enhancing the aesthetic value of the watershed lakes and streams.

---

**Building Partnerships**

With the Lake Olathe Project, the City took the initiative to get engaged in source water protection, and engaged citizen participation in the process of developing a comprehensive watershed protection plan for Lake Olathe. The initiative helped build a number partnerships with local, state, and federal agencies, and it enabled the City's water department to reach out to citizens and the community of Olathe in a positive way because the project demonstrated a commitment to water customers beyond what is merely required by regulations.

**Figure 7.2: Building Partnerships**

In order to develop a comprehensive watershed protection plan that is responsive to community needs, a citizen advisory board was formed to engage public participation from local residents and other stakeholders in the decision-making process. From January 2003 through June 2004, members of the Lake Olathe Watershed Protection Advisory Board (LOWPAB) worked to develop recommendations for ensuring the long-term health of the Lake Olathe Watershed. Recommendations from the LOWPAB included the following:

- Designating the Lake Olathe watershed as a Watershed Protection Area.
- Incorporating the Lake Olathe Watershed Protection Area into key policy documents and plans, such as the comprehensive land use plan and unified development ordinance.
- Developing a source water protection ordinance to manage land use activities within the watershed to protect public health, ensure the availability of safe drinking water, and prevent the degradation of water quality in Lake Olathe and Cedar Lake.
- Developing an ordinance to require stream buffers to filter pollutants and reduce flooding, as well as to maintain the integrity and aesthetics of the streams in the Lake Olathe Watershed.
- Developing site design practices that reduce impervious cover within the watershed protection area to ensure that development is given full opportunity to be designed in ways that minimize negative impacts of urban stormwater in the Lake Olathe watershed.
- Continuing to enhance City standards for erosion & sediment control to identify standards and practices to reduce the amount of sediment and other pollutants in runoff from construction sites.
- Continuing to enhance City standards for stormwater management facilities and management practices in order to help minimize the extent to which hydrological conditions are altered during development.
- Developing policies and practices for city operations that address pollution prevention to reduce the impact that the City's own activities will have on the health of the Lake Olathe Watershed.
- Developing a public outreach campaign to increase awareness about water quality concerns in Lake Olathe, and to educate citizens about their role in protecting the overall health of the Lake Olathe Watershed.
- Developing indicators to measure progress toward achieving goals by implementing a monitoring program to evaluate changes over time in the rate of eutrophication in Cedar Lake and Lake Olathe.
- Developing a strategy to implement "retrofitting" of current development to incorporate best management practices for stormwater management in order to mitigate the negative water quality impacts associated with past design and construction.
- Developing a restoration strategy for Cedar Lake and streams within the watershed to reduce stream bed and bank erosion and overall pollutant loads delivered to Lake Olathe.

- Improving management of agricultural land use activities to minimize pollutants in stormwater runoff from agricultural lands.

- Mitigating the impacts of quarry operations to improve the integrity of the surrounding streams by reducing the amount of sediment and other pollutants in stormwater runoff from quarrying activities.

Utility managers at the City of Olathe took some risk by moving forward with this effort. The City is generally considered to be a pro-growth community, and efforts to slow development are often not politically popular. City leaders contribute the success of the initiative to three things:

1. First, strong leadership by utility managers. Had utility managers not made this a priority, none of the other groups in the community would have made this issue a priority and no action would have been taken as a result.

2. Second, broad participation in the process by community leaders. The people selected to be on the LOWPAB advisory committee included some of the most influential leaders in the community. By engaging senior leaders from the community in the process, the recommendations were stronger and more credible to the community.

3. Third, collaboration with a wide range of organizations. Utility managers at the City of Olathe initiated collaboration with a wide range of organizations including: the U.S. geological society, the Kansas Department of Heath and Environment, the Environmental Protection Agency, Johnson County, Kansas State University, the Kansas Biological Survey, and regional planners from the area's metropolitan planning organization (Mid America Regional Council), the Olathe City Council, the Olathe Planning Commission, and others. By being the ones to initiate the effort, utility managers at the City of Olathe were perceived to be leaders among a wide range of organizations that are committed to the environment.

## COMMUNICATION AND INVESTMENT

During the fall of 2000, the City of Olathe Municipal Services Department began administering quarterly customer surveys to assess the quality of service delivery and the effectiveness of communication efforts. After compiling more than 10 quarters of data over a two and half year period, utility managers were able to demonstrate that communication from the Department was having a positive impact on customer satisfaction and awareness of service related issues. Specifically, the survey results showed increases in awareness and satisfaction each time the utility distributed publications, such as the Consumer Confidence Report (CCR) and information about conservation techniques.

Utility managers used the results of the quarterly survey research to demonstrate the need to have a regular publication from the City about utility issues. In the fall of 2003, utility managers were given space for two pages that were added to the City's newsletter (Figure 7.3). Utility managers at the City decided to use the opportunity to support the City's mission of being an environmental leader by naming the new section of the City's Newsletter the *Olathe Earth News*.

Although basic utility information is contained in each issue, the department emphasizes environmental issues in each publication.

> **Communication**
>
> During the fall of 2003, the City of Olathe added a 2-page section called the Olathe Earth News to the City's newsletter. The section focuses on environmental initiatives being undertaken by the city along with messages about ways residents can help protect the environment. Before the Olathe Earth News was started, 52% of the City's residents indicated they read the City's newsletter on a regular basis. A year after the Olathe Earth News feature was added, 68% of the City's residents indicated they read the newsletter on a regular basis.

**Figure 7.3: Communication**

**Investing in "Green" Facilities**

In Dec 2003, the City of Olathe opened its new Municipal Services Center, which serves as the central facility for the City's utility and customer service operations. The facility was the first public building in the State of Kansas to be certified as a "green" building and cost more than $30 million to complete. Although the environmentally friendly structure cost significantly more than traditional construction, senior managers at the utility felt it was important to make the investment given the organization's commitment to environmental leadership. "This building shows that our city is really committed to the environment," says the current Municipal Services Director, Don Siefert.

The City carefully managed the public "roll-out" of the new facility so that the media attention was focused on the environmental benefits of the new structure and City's commitment to being an environmental leader. City officials pointed to survey results that showed citizen support of the new facility and asked community leaders who were involved in the LOWPAB Advisory Committee to be available to make positive comments about the new facility. As result, media coverage of the new facility was very positive and included two front page stories about the new facilities.

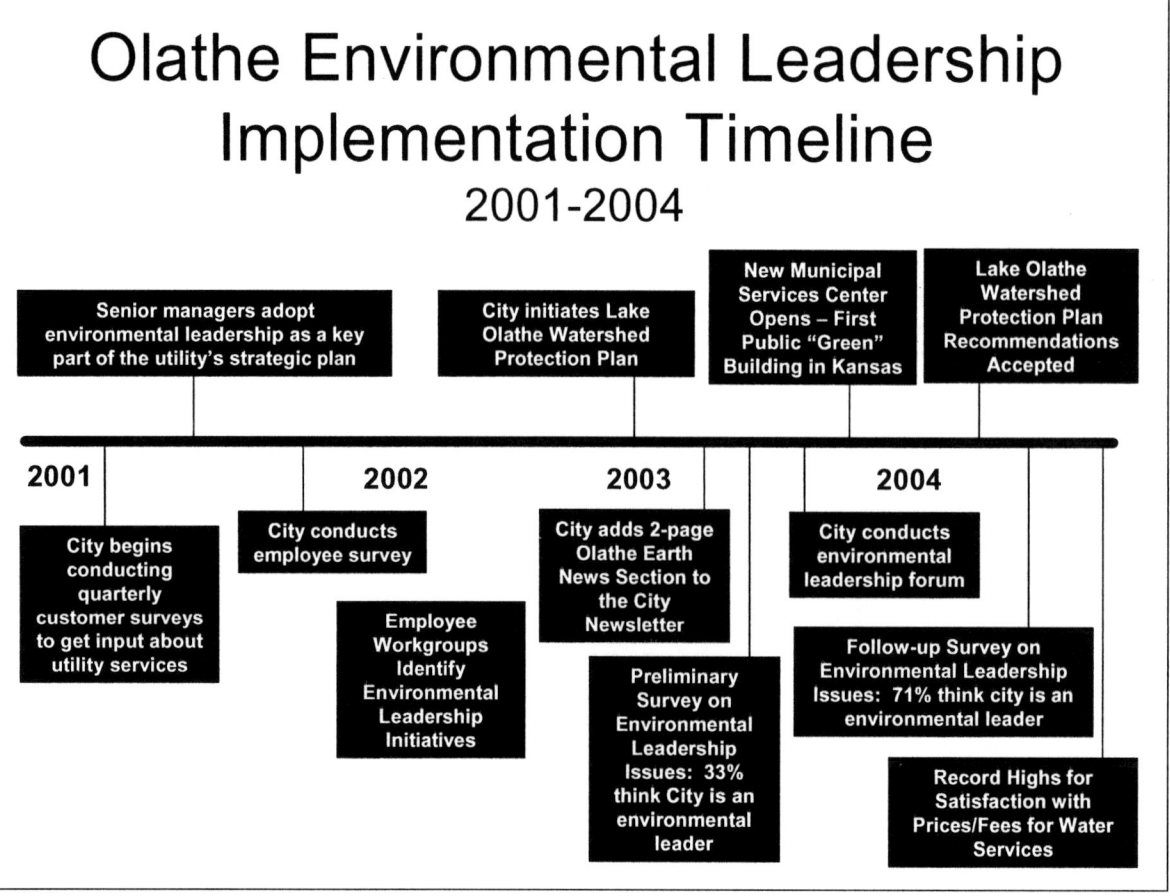

**Figure 7.4: Implementation Timeline**

## IMPACT OF THE CITY OF OLATHE'S EFFORTS

The results of the City's efforts have been impressive (see Timeline above in Figure 7.4 for a summary). The City of Olathe has been able to raise water utility rates without opposition for four consecutive years. Overall satisfaction with the City's water services are at an all time high, including overall satisfaction with the value of services received, even though rates are higher.

In order to evaluate the impact of the City's efforts, the research team administered a survey to a random sample of 400 Olathe residents in October 2003 and again in November 2004. The 2003 survey was administered to provide a baseline for objectively assessing the impact that the City's environmental leadership initiatives had. The results of each survey have a precision of at least +/-5% at the 95% level of confidence. Major findings are shown below.

**Finding 1: The number of residents in the city who thought it was important for the City to be an environmental leader increased significantly.**

As the chart on the next page (Figure 7.5) shows, the percentage of Olathe residents who thought it was "very important" for their water utility to be an environmental leader increased from 58% to 80% as a result of the City's Environmental Leadership Initiatives.

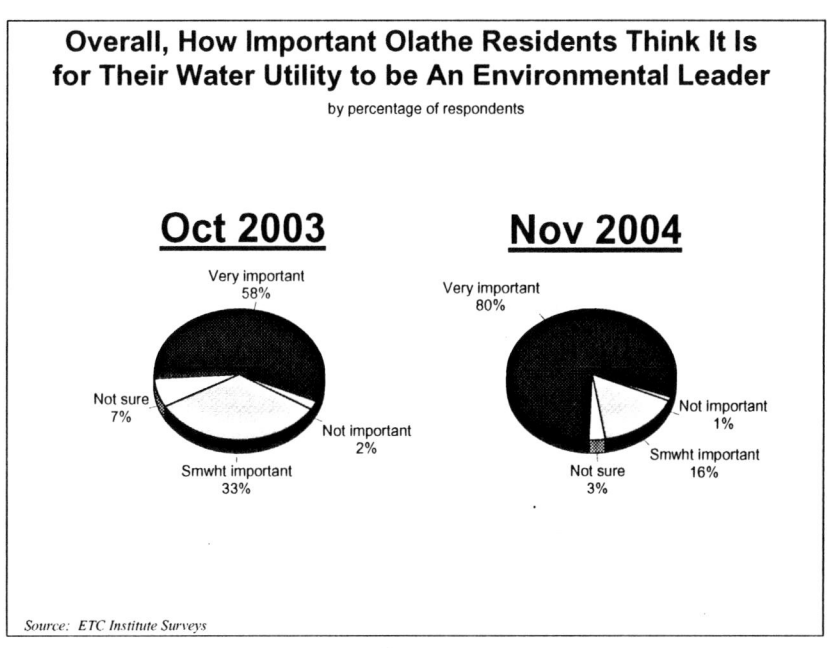

**Figure 7.5: Importance of Environmental Leadership**

**Finding 2: The percentage of residents who thought the City was an environmental leader increased dramatically in just 13 months.**

As the chart below shows, the percentage of Olathe residents who thought their water utility is an environmental leader increased from 33% in October 2003 to 71% in November 2004 as a result of the City's Environmental Leadership Initiatives (Figure 7.6).

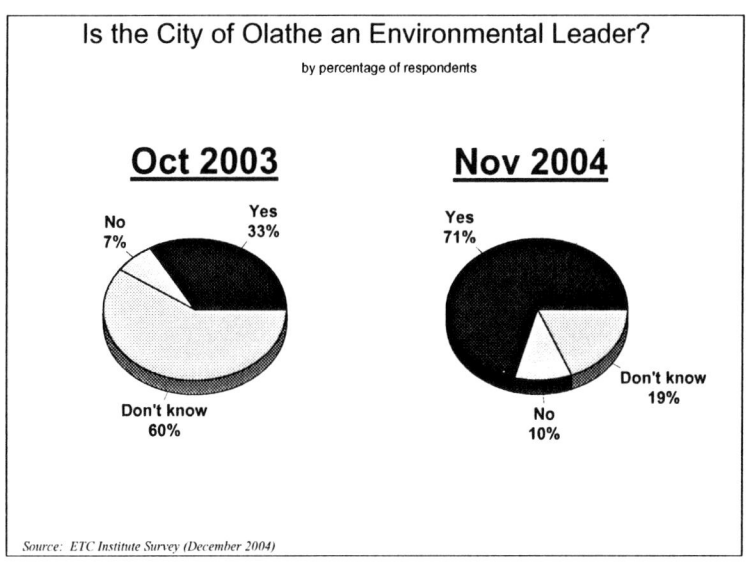

**Figure 7.6: Perceptions of Environmental Leadership**

**Finding 3: The percentage of residents who were informed about environmental issues increased significantly**

The number of Olathe residents who thought they were "very informed" about environmental issues nearly doubled while the number who were "not informed" was reduced by more than half in just 13 months. See Figure 7.7 below.

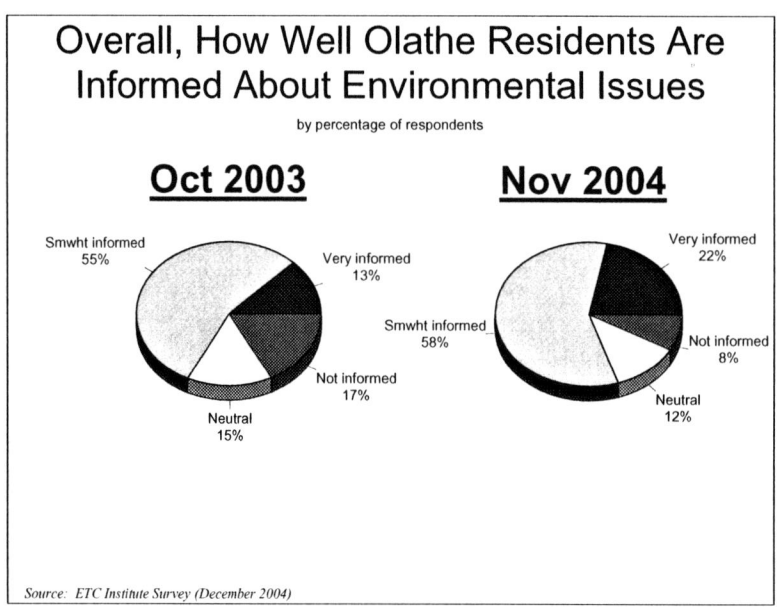

Figure 7.7: Awareness of Environmental Issues

**Finding 4: Overall Satisfaction with Fees and Prices Reach New Highs.**

Overall satisfaction with the prices and fees paid by Olathe residents for water services reached a four-year high during the fall of 2004 even though the City had increased water rates annually since 2000. Given the City's strong emphasis on environmental leadership during the past year, the higher levels of satisfaction with water services may be related to an increase in the perceived value of the services provided

**CONCLUSIONS**

The actions by the City of Olathe Municipal Services Department between 2001 and 2004 may serve as a model for many water utilities that want to be seen as environmental leaders in their community. By taking the initiative to lead environmentally responsible action before environmental issues presented a crisis in the community, the City of Olathe was able to dramatically change the perception of residents in a relatively short period of time. In less than two years, the number of residents who thought the City was an environmental leader increased from 33% to 71%. Overall satisfaction with the utility reached new highs, and city leaders made unprecedented investments in environmentally friendly infrastructure.

The research team believes that the Olathe model demonstrates the need for water utilities to take action before environmental issues become a crisis. In communities where environmental concerns are already a crisis, such actions by a water utility are more likely to be perceived as compliance rather than leadership. Utilities that make environmental leadership a priority before there is a crisis in their community are much more likely to achieve positive results with less effort than utilities that wait for a crisis. By being proactive on environmental issues, water utilities are also likely to generate "good will" in the community which will make it easier for the utility to collaborate with environmental interests in the future.

# APPENDIX A
# ORGANIZING AN ENVIRONMENTAL LEADERSHIP ADVISORY BOARD

# ORGANIZING AN ENVIRONMENTAL LEADERSHIP ADVISORY BOARD

To assist in the creation of the strategies presented in this report, recruit representatives of the following groups and organizations to serve on your Environmental Leadership Advisory Board.

- Boards of education.
- Community development/planning agencies (city and county).
- Non-government organizations.
  Examples: home owner associations, chambers of commerce, and industry associations.
- City government departments, including those with water regulatory authority and those that do not regulate water quality.
- County government departments, including those with water regulatory authority and those that do not regulate water quality.
- State government agencies, including those with water regulatory authority and those that do not regulate water quality.
- Federal government agencies, including those with water regulatory authority and those that do not regulate water quality.
- Relevant elected or appointed officials (city, county, state).
- Community leaders.
- Agricultural, business/commercial, and industrial water consumers.
- Residential water consumers.
- Engineers (consulting, civil).
- Environmental activist/advocacy groups; watershed interest groups.
- Public health officials and other health sector representatives, including physician and dentist organizations.
- Leisure activity advocates.
- Recreational water users.
- Private landowners.
- Real estate developers, builders, and landscape contractors.
- Scientists (biological, social, behavioral).
- Tribal leaders.
- Water utility managers and operations personnel.
- Community service organizations.
- Water management district (where it exists).
- League of Women Voters.

# APPENDIX B
# ENVIRONMENTAL LEADERSHIP FORUM SUMMARIES

# ENVIRONMENTAL LEADERSHIP FORUM SUMMARIES

The research team conducted environmental leadership forums in each of the participating utility markets. These forums were used to engage community leaders in the study process and provide an opportunity for the participating utilities to demonstrate their commitment to becoming environmental leaders.

Not all of the ideas from the forums were incorporated into the strategies that have been described in this report. To ensure that these ideas are not lost, a summary of each of the forums is provided in this appendix.

Water utility managers are encouraged to review the content of this appendix when planning events that will engage community leaders in discussions about environmental issues. In addition, water utility managers are encouraged to use the processes described in this report to plan and conduct their own Environmental Leadership Forums. These meetings provide an excellent opportunity for water utilities to begin establishing partnerships with other organizations in the community.

# AwwaRF Environmental Leadership Forum Summary
# Fort Lauderdale, Florida
# October 17, 2003

## OVERVIEW

On Friday, October 17, 2003, ETC Institute facilitated an Environmental Leadership Forum for Fort Lauderdale, Florida, sponsored by the American Water Works Association Research Foundation. The forum was held at the Doubletree Galleria hotel, 2670 Sunrise Boulevard in Fort Lauderdale, Florida.

The purpose of the forum was two fold: 1) to present the results of a recent survey and other research showing how Fort Lauderdale compares to other communities across the United States on a wide range of issues related to drinking water, and 2) to identify and discuss opportunities and initiatives to maintain and enhance Fort Lauderdale's position as an environmental leader in the 21$^{st}$ Century.

Stakeholders from public and private sectors from the Fort Lauderdale region were invited to the forum. Forty-two (42) stakeholders attended.

Chris Tatham, Vice President of ETC Institute, provided an overview and the purpose of the project. Dr. Robert Cicerone of ETC Institute, provided selected findings from Phase 2 of the project: The Results of a Regional and National Survey on Environmental Leadership.

David Gluckman, from Tallahassee, an expert on environmental law, was the guest speaker.

## METHODOLOGY

During the afternoon session of each Environmental Leadership Forum, participants were formed into small break-out groups of 10 to 15 people. The members of each break-out group were asked to write down their personal suggestions for strengthening the area's position as an environmental leader. They were then asked to share with the group the one strategy from their own list that they judged to be the most important. The facilitator recorded the most important strategy from each participant on a flipchart. After all members of a break-out group described their most important strategy, like strategies were combined and the unique strategies were then ranked in order of their importance.

The strategy ranked most important by the break-out group members was then analyzed to determine why the strategy was important (Topic 1), the organization(s) and/or person(s) who should be involved, including the person or organization that should lead the strategy (Topic 2), the specific actions that the local area and organizations should take to implement the strategy (Topic 3) and finally the identification of obstacles or barriers that might reduce the success of this strategy and what could be done to avoid or minimize the effects of these barriers or obstacle (Topic 4). If time permitted, this analysis was done for other strategies ranked highly important by the group members.

For Fort Lauderdale, the most important strategies for strengthening the area's position of environmental leadership are provided on the following pages.

**Strategy 1:** **Ecological Democratization Process**

    Topic 1 – Why the strategy is important?  When should it start?
- All the stakeholders are there to create buy in.

    Topic 2 – Which people and/or organizations should be involved?
- The utility director using CAP process, plus a steering committee.

    Topic 3 – What actions should Fort Lauderdale and the other organizations take to implement the strategy? Commission approval and buy in from the community.

    Topic 4 – What would be the obstacles to the plan?
- Confrontation, conflict, fear of change.  The education process for the public would need to include greatly simplifying a very technical subject.

**Strategy 2:** **Greater Public Education**

    Topic 1 – Why the strategy is important? When should it start?
- There is a basic mistrust of government by the public.

    Topic 2 – Which people and/or organizations should be involved?
- A public information officer.

    Topic 3 – What actions should Fort Lauderdale and the other organizations take to implement the strategy?
- Foster a better relationship with the media.
- Look for "best practices" within the community and promote them through the schools and the community.
- Utilize many tools: CCR (better review this document for the education effort), water bill inserts, develop Water Department icon (brand image)

    Topic 4 – What would be the obstacles to the plan?
- Money
- Relationship with the media.

**Strategy 3:** **Better communication at the leadership level.**

    Topic 1 – Why the strategy is important? When should it start?
- We need to improve the disconnect or gap between the county and municipalities – reduce parochialism.

    Topic 2 – Which people and/or organizations should be involved?
- The County Commissioner who is most environmentally aware.
- Utility director.
- Fort Lauderdale Commissioners.

Topic 3 – What actions should Fort Lauderdale and the other organizations take to implement the strategy?
- There should be a meeting of Utility Directors and County Commissioners to determine the best steps to take. Utilize the Water Advisory Board to provide direction for better communication.

Topic 4 – What would be the obstacles to the plan?
- Getting past parochial mind sets.
- There is no controlling County Commissioners and their communication efforts.

**Strategy 4:** **Incentive reward program for environmentally responsible behavior.**

Topic 1 – Why the strategy is important? When should it start?
- You can get the developers and the public to buy in. It would get results – in some cases, long term results.

Topic 2 – Which people and/or organizations should be involved?
- It should be lead by the City Manager.
- A task force needs to be established.
- Major research is needed to identify meaningful rewards (perhaps outside research). It would have to be the foundation of the eventual reward/incentive program.

Topic 3 – What actions should Fort Lauderdale and the other organizations take to implement the strategy?
- Establish a task force to develop the program.
- Conduct research to identify meaningful incentives/rewards.
- Conduct workshops with public and large entities like developers.
- Get Commission approval.
- Create ordinances.

Topic 4 – What would be the obstacles to the plan?
- It's a big effort and a lot of time would be required.
- Funding.
- Time and the right people.

**Strategy 5:** **More public education about utility operations, regulatory requirements, and environmental leadership levels.**

Topic 1 – Why the strategy is important? When should it start?
- The education effort should be on-going and continuous, not sporadic or just when a crisis occurs
- Pre-emptive – utilities should take the initiative, not only in terms of what they are doing, but what they are planning.

- Communication should be framed on what you know about the population – do market research if necessary.
- Communication is important to the County Commission, to garner support.

Topic 2 – Which people and/or organizations should be involved?
- County School Board
- Community Development Department – City and County.
- South Florida Water Management District
- Non-governmental organizations
- League of Women voters
- Form collaborative associations with organizations as appropriate.

Topic 3 – What actions should Fort Lauderdale and the other organizations take to implement the strategy?
- Assess what's expected;
  -Potable water issues
  -Stormwater issues
  -Wastewater/Sewer issues
- Analyze and evaluate using 3 disciplines; biological, social sciences and behavioral sciences.
- Identify exemplars re:
  -Communicating with the general public
  -Adapt practices to each situation, also within city organization
  -Look outside the City – formulate
  -Communications to various populations as public education.

Topic 4 – What would be the obstacles to the plan?
- A lack of adequate funding:
  -Reflects on lack of position in department (who will do it)
  -Impacts media to use, such as television
  -Political resistance
  -Low priority.

**Strategy 6: Public Outreach**

Topic 1 – Why the strategy is important? When should it start?
- These are the people who use the resources and pay the bills – they need to be educated.
- Start young.
- They need to understand they are the resource stewards and how to be resource stewards.

Topic 2 – Which people and/or organizations should be involved?
- Water Utility, Politicians, all levels of government including city staff, environmental groups/organizations, business and industry, school board, civic organizations, press, residential consumers.

Topic 3 – What actions should Fort Lauderdale and the other organizations take to implement the strategy?
- Advertise on TV, the best way to reach the public.
- Hold community events.
- Train people, especially school kids, to test and monitor water quality, with an understanding of the results.
- Video tours.
- Inform the public about what is being done.
- Speaking bureau.
- Publications.
- Involve children - they will get their parents involved
- Curriculum in school.
- Reach out to churches
- Get money from foundations.
- "Conservation is your patriotic duty"

Topic 4 – What would be the obstacles to the plan?
- It is tough getting public participation.
- On-site tours no longer possible due to increased security since 9/11.
- Employees need to take the time to do public speaking.
- Limited resources.
- Too many dollars diverted to security.

**Strategy 7: Water Resource Management**

Topic 1 – Why the strategy is important? When should it start?
- Sustain resources.
- Avoid waste – waste equals money.
- Conserve water.

Topic 2 – Which people and/or organizations should be involved?
- Water Resource Management (County Department of Environmental Planning).
- Neighboring utilities and drainage districts.

Topic 3 – What actions should Fort Lauderdale and the other organizations take to implement the strategy?
- Create more wet season water storage.
- Take the lead – well fields recharged.
- Stormwater discharge program – treatment and usage
- Better use of freshwater loss to the ocean.
- Redirect downstream water flows
- Reuse water.

Topic 4 – What would be the obstacles to the plan?
- There is no space in Fort Lauderdale to store stormwater.

- High cost.
- Regulations.
- Urban related problems.

**Strategy 8:** **Regionalization**

Topic 1 – Why the strategy is important? When should it start?
- We all live downstream from someone.
- Economies of scale.
- Water is not a local issue.

Topic 2 – Which people and/or organizations should be involved?
- Local utilities.
- Public.
- State legislature
- Everybody.

Topic 3 – What actions should Fort Lauderdale and the other organizations take to implement the strategy?
- State water supply
- Better coordination with county on water management plan and new urban master plan.

Topic 4 – What would be the obstacles to the plan?
- Developers

# AwwaRF Environmental Leadership Forum Summary
## Kansas City Metropolitan Area
## December 2, 2003

## OVERVIEW

On Tuesday, December 2, 2003, ETC Institute facilitated an Environmental Leadership Forum for the Metropolitan Kansas City area, sponsored by the American Water Works Association Research Foundation. The forum was held at the Olathe Municipal Services Center (the first "green" building in Kansas), at 1385 Robinson, Olathe, KS 66061.

The purpose of the forum was two fold: 1) to present the results of a recent survey and other research showing how the Metropolitan Kansas City area compare to other communities across the United States on a wide range of issues related to drinking water, and 2) to identify and discuss opportunities and initiatives to maintain and enhance the Metropolitan Kansas City area's position as an environmental leader in the 21$^{st}$ Century.

Stakeholders from public and private sectors from the Metropolitan Kansas City Area were invited to the forum. Fifty-five (55) stakeholders attended.

Chris Tatham, Vice President of ETC Institute, provided an overview and the purpose of the project. Dr. Robert Cicerone of ETC Institute, provided selected findings from Phase 2 of the project: The Results of a Regional and National Survey on Environmental Leadership.

Steve Maxwell, a national authority on trends in the water industry from Boulder, Colorado, was the featured speaker, discussing applications for environmental leadership in the Kansas City Region.

## METHODOLOGY

During the afternoon session of each Environmental Leadership Forum, participants were formed into small break-out groups of 10 to 15 people. The members of each break-out group were asked to write down their personal suggestions for strengthening the area's position as an environmental leader. They were then asked to share with the group the one strategy from their own list that they judged to be the most important. The facilitator recorded the most important strategy from each participant on a flipchart. After all members of a break-out group described their most important strategy, like strategies were combined and the unique strategies were then ranked in order of their importance.

The strategy ranked most important by the break-out group members was then analyzed to determine why the strategy was important (Topic 1), the organization(s) and/or person(s) who should be involved, including the person or organization that should lead the strategy (Topic 2), the specific actions that the local area and organizations should take to implement the strategy (Topic 3) and finally the identification of obstacles or barriers that might reduce the success of this strategy and what could be done to avoid or minimize the effects of these barriers or obstacle (Topic 4). If time permitted, this analysis was done for other strategies ranked highly important by the group members.

For the Metropolitan Kansas City area, the most important strategies for strengthening the area's position of environmental leadership are provided on the following pages.

**Strategy 1:   Unify all water issues**

Topic 1 – Why the strategy is important?  When should it start?
- The strategy to unify water issues is important to address before it becomes a major dilemma in the area, and while current growth can be influenced. It should have started "yesterday" – it's important to start it now.

Topic 2 – Which people and/or organizations should be involved?
- The Mid-America Regional Council should be the administrator of a unified environmental approach to water issues but they would have to be approached by an organization(s) who can contribute funding and leadership. Organizations and people mentioned were the School Districts, Elected officials, Water Utilities, Wastewater Utilities, Regulatory Agencies, State officials, grass roots groups, and those with relevant expertise.

Topic 3 – What actions should the Kansas City Metropolitan area and the other organizations take to implement the strategy?
- Someone should study the issue and identify the environment issues that need attention. One of the Water stakeholder organizations should fund the study and a campaign should be started to unify all of the stakeholders.

Topic 4 – What would be the obstacles to the plan?
- Public perception – we need to unify our goals with other organizations.
- Agriculture community is a primary contributor to environmental problems with water and their cooperation would be a challenge.

**Strategy 2:   Educate the public.**

Topic 1 – Why the strategy is important?  When should it start?
- An educational campaign should begin now, to assure that future generations don't suffer from the actions of this generation.

Topic 2 – Which people and/or organizations should be involved?
- Cities, municipalities, engineers, MARC, grass roots organizations, water utilities, politicians. We should be consulting with organizations that have already done environmental educating.

Topic 3 – What actions should the Metropolitan Kansas City area and the other organizations take to implement the strategy?
- Bring private, public and non-profit organizations together to help guide the development of policy.

Topic 4 – What would be the obstacles to the plan?
- Territorial boundaries – actual and political.

**Strategy 3:** **City leaders need to make the environment a priority and set an example by their commitment.**

Topic 1 – Why the strategy is important? When should it start?
- Our resources are limited. An example of good stewardship needs to be set by our leaders for the public to follow. It will cost less to address it now, before it becomes a crisis. A long term plan needs to be developed.

Topic 2 – Which people and/or organizations should be involved?
- Chamber of Commerce, Service clubs, League of Women Voters, State Legislature, Agricultural community, City leaders and elected staff.

Topic 3 – What actions should the Metropolitan Kansas City area and the other organizations take to implement the strategy?
- Zoning and planning ordinances that promote "green" watershed protection. Coordinate activities within the city and other entities (counties) to promote environmental leadership. Model environmental leadership in maintaining city properties (salt storage, vehicle maintenance, erosion control).

Topic 4 – What would be the obstacles to the plan?
- The magnitude of who impacts/coordinates/controls.
- Information to policy makers.

**Strategy 4:** **Link leadership of organizations; build teamwork/communication/ coordination.**

Topic 1 – Why the strategy is important? When should it start?
- It is a means of educating the public.
- It is a time savings to prepare and implement actions.
- It reduces conflict.

Topic 2 – Which people and/or organizations should be involved?
- County, City, Agriculture, Schools, Neighbor, Chamber, Service organizations, Professional Associations, Developers/Real Estate, Lawn & Garden Businesses, Golfers.

Topic 3 – What actions should the Metropolitan Kansas City area and the other organizations take to implement the strategy?
- Individual, grassroots education
- The City should survey groups with questions about drinking water, use of terminology, etc.
- Citizen priorities
- Elevate environmental leadership priorities
- Develop favorable public policy.

<u>Topic 4 – What would be the obstacles to the plan?</u>
- Disagreement/lack of consensus
- Apathy
- Voter turnout
- Identify all stakeholders to assure equity and to identify diversity issues.
- Address individual needs vs. common good.
- Money

# AwwaRF Environmental Leadership Forum
# San Diego, California
# November 21, 2003

## OVERVIEW

On Friday, November 21, 2003, ETC Institute facilitated an Environmental Leadership Forum for the San Diego region of California, sponsored by the American Water Works Association Research Foundation. The forum was held at the Embassy Suites, San Diego Bay, 601 Pacific Highway, San Diego, California.

The purpose of the forum was two fold: 1) to present the results of a recent survey and other research showing how San Diego compares to other communities across the United States on a wide range of issues related to drinking water, and 2) to identify and discuss opportunities and initiatives to maintain and enhance San Diego's position as an environmental leader in the $21^{st}$ Century.

Stakeholders from public and private sectors from the San Diego region were invited to the forum. Thirty four (34) stakeholders attended.

Chris Tatham, Vice President of ETC Institute, provided an overview and the purpose of the project. Dr. Robert Cicerone of ETC Institute, provided selected findings from Phase 2 of the project: The Results of a Regional and National Survey on Environmental Leadership.

Peter Silva, Vice Chair of the State Water Resources Control Board, was the featured speaker. He addressed applications for this project in the San Diego region.

## METHODOLOGY

During the afternoon session of each Environmental Leadership Forum, participants were formed into small break-out groups of 10 to 15 people. The members of each break-out group were asked to write down their personal suggestions for strengthening the area's position as an environmental leader. They were then asked to share with the group the one strategy from their own list that they judged to be the most important. The facilitator recorded the most important strategy from each participant on a flipchart. After all members of a break-out group described their most important strategy, like strategies were combined and the unique strategies were then ranked in order of their importance.

The strategy ranked most important by the break-out group members was then analyzed to determine why the strategy was important (Topic 1), the organization(s) and/or person(s) who should be involved, including the person or organization that should lead the strategy (Topic 2), the specific actions that the local area and organizations should take to implement the strategy (Topic 3) and finally the identification of obstacles or barriers that might reduce the success of this strategy and what could be done to avoid or minimize the effects of these barriers or obstacle (Topic 4). If time permitted, this analysis was done for other strategies ranked highly important by the group members.

For San Diego, the most important strategies for strengthening the area's position of environmental leadership are provided on the following pages.

**Strategy 1:** **Launch a public relations campaign to unify the environmental message throughout the San Diego region.**

Topic 1 – Why the strategy is important? When should it start?
- Like Shell Oil and Sea World have done in promoting themselves as environmental leaders, San Diego should promote the regional identity as an environmental leader. There should be one cohesive message that unifies the city and the surrounding region with a strong environmental image and water management should be a prominent player.

- Water management can be a major catalyst in accomplishing this goal if they lead by example:
  - Developing a strategic plan for water supply planning
  - Water conservation
  - Watershed management
  - Promote protecting the local water supplies
  - Provide cost savings with incentive programs
  - Educate the public in many ways, including poster contests, etc.

- A public relations campaign should start <u>now</u>.

Topic 2 – Which people and/or organizations should be involved?
- The City Manager should lead the public relations campaign with the Water Department and staff being the catalysts. The campaign needs to sound like <u>one voice,</u> including CIP, conservations organizations, legislation, etc.

Topic 3 – What actions should San Diego and the other organizations take to implement the strategy?
- Taking a <u>regional</u> perspective is very important. The CWA may take the lead for common issues, but what should be avoided is the duplication of effort that will result if each organization makes an individual attempt at their own environment-related PR campaign. San Diego (the region) should unify on this issue to have the greatest impact.

Topic 4 – What would be the obstacles to the plan?
- The greatest obstacle will be the unifying as one voice, forgetting territorial issues, relationship issues, cost.

**Strategy 2:** **Develop a water education program to gain appreciation for supply and demand challenges in the region.**

Topic 1 – Why the strategy is important? When should it start?
- It is a long overdue approach to water problems in the region and should be implemented now.

Topic 2 – Which people and/or organizations should be involved?
- Building Industry Associations, NAIOP, Quality of Life Coalition, Water

Conservation Authority, Landscape Contractors, Engineering Community, San Diego River Conservancy, SDCWA, SD County Department of Education, SDSU, San Diego City Schools, Endangered Habitat League, Seniors, Indian Tribes.

Topic 3 – What actions should San Diego and the other organizations take to implement the strategy?
- Create a water story video production.
- Provide a demonstration for water conservation conversion project.
- Conduct workshops/conferences/forums.
- Institute a traveling classroom.
- Contact speaker's bureaus.

Topic 4 – What would be the obstacles to the plan?
- Elected official acceptance.
- Identifying a funding stream.
- Availability of staff.
- Community cooperation to create a unified vision.

**Strategy 3:** **Conserve water to the maximum extent possible. Allocate resources as needed to work on environmental issues. Determine what is important to the public. Encourage water conservation pricing. Offset impact of population growth encouraged.**

Topic 1 – Why the strategy is important? When should it start?
- We need to allocate resources to conserve water, which will require addressing regulatory issues. If we minimize negative impacts, we will improve our environment.

Topic 2 – Which people and/or organizations should be involved?
- Include the entire region and contact environmental groups to identify the important environmental issues. Partnership with groups who may have similar interests, like developers, other agencies, and regulatory groups.

Topic 3 – What actions should San Diego and the other organizations take to implement the strategy?
- Solicit opinions from the public and stakeholder groups.
- Reach agreements with governing bodies about the issues to be addressed. Identify potential resources/partnerships. Agree on a plan to implement.
- Allocate available resources.
- Inform the public of the issue and what's being done about it.

Topic 4 – What would be the obstacles to the plan?
- Governing bodies with competing priorities.
- Rate payers may view other issues besides this one as more important.
- Regulatory issues.
- Insufficient resources.

**Strategy 4:** Water Utility managers need to take the initiative to go beyond just compliance.

Topic 1 – Why the strategy is important? When should it start?
- Address regulatory and compliance.
- Minimize negative impact of utility operations on environment.
- Help DWW properly allocate resources.
- Enhance environment – restore ground water supplies.

Topic 2 – Which people and/or organizations should be involved?
- Regulatory
- Watershed – environmental advocates and activists
- Community groups
- Pool/combine resources from other DWW and other relevant groups like stormwater agencies and real estate development groups.

Topic 3 – What actions should San Diego and the other organizations take to implement the strategy?
- ID the users to be addressed on internal assessment, solicit ideas from the public, and solicit from multiple groups.
- RA to relevant governing bodies regarding issues to be addressed and plans to be implemented regarding sources and resources.
- Allocate resources to address issues applying to governing bodies – access internal revenues, ID external sources of funds, combine resource3s to other organizations – partnering.
- Inform your public about the issues – the reason it's important, the basic plan, and funding.

Topic 4 – What would be the obstacles to the plan?
- The governing body establishes a higher priority for another issue.

# APPENDIX C
# HOW TO PLAN AN ENVIRONMENTAL LEADERSHIP FORUM IN YOUR COMMUNITY

# HOW TO PLAN AN ENVIRONMENTAL LEADERSHIP FORUM IN YOUR COMMUNITY

Drinking water utility managers can use the following steps to conduct an environmental leadership forum in their service area, watershed, or region. This appendix contains a step-by-step description of the process for conducting an Environmental Leadership forum. Samples of invitation letters and agenda are provided at the end of this Appendix.

**Step 1:** Create a project management team to lead the development, implementation, and evaluation of the environmental leadership forum; include personnel from:
- All levels of the utility's work force (management, non-management, exempt, non-exempt).
- Unions that represent utility employees.
- All functional areas of the utility.

**Step 2:** As appropriate, revise the general plan contained in this appendix to create a plan that is detailed and specific to your utility.

**Step 3:** Reach agreement with the body that governs the utility about:
- The objectives to be achieved by an environmental leadership forum (what should be accomplished).
- Plan to achieve the objectives.

**Step 4:** If not previously done, form an Environmental Leadership Advisory Board to give advice on environmental leadership issues. Organizations from which to recruit people to serve on this Board are listed in Appendix A of this report. Consult with the Advisory Board during the planning of an environmental leadership forum.

**Step 5:** Identify which of the utility's strategic objectives could be supported by an environmental leadership forum.

**Step 6:** For each strategic objective listed in Step 5:
- Determine what observable outcomes of an environmental leadership forum would show that the forum supported the objective.
- Decide which measures will be used to assess the outcomes of a forum.

**Step 7:** Given the strategic objectives (Step 5) and expected outcomes (Step 6), identify the group(s) whose attendance would most likely enable the expected outcomes to be achieved. (E.g., customers (business/commercial, industrial, residential) and stakeholders who are not customers.)

**Step 8:** Create a list of people to invite from the groups identified in Step 7.

**Step 9:** Create a topic agenda that will achieve the expected outcomes of the forum described in Step 6 with the groups identified in Step 7.

**Step 10:** Given the group(s) to be invited (from Step 7) and the topic agenda (from Step 9), select activities (e.g., presentation, movie, panel discussion, role plays, etc.) and media (e.g., printed handout, power point presentation, slides, overhead transparencies, etc.) to be used.

**Step 11**: Buy and/or create the materials selected in Step 10.

**Step 12**: Decide on these administrative matters:
- Day, date, and time.
- Location for the forum (and schedule the meeting room).
- Whether to provide food and beverage service during the forum (and arrange for food and beverage service).

**Step 13:** Develop the invitation; include the following content:
- Reason(s) the forum is being held.
- How the forum will benefit the region, watershed, utility service area, utility, general public, water users, and/or the invitee.
- Where the forum will be held, day/date, and times.
- (optional) Food and/or beverages will be supplied.
- If there will be a guest speaker, the speaker's name, credentials, and topic on which they will speak.
- How to notify the utility whether they will attend – and a deadline by which to respond to the invitation.
- Expression of hope the invitee will attend.

**Step 14**: Mail the invitations about 30-40 days before the scheduled date.

**Step 15:** About 7-10 business days after the invitation is mailed, telephone those invitees who have not yet responded to the invitation; offer to answer their questions about the forum; ask whether they will attend; record the names of those who say they will attend and send them a written confirmation along with directions to the meeting location.

**Step 16:** Practice conducting the forum.

**Step 17:** One or two days before the forum is held, telephone all those who said they would attend to remind them.

**Step 18:** Conduct the forum.

**Step 19**: Collect data relevant to the measures selected in Step 6 to assess the outcomes of a forum.

**Step 20**: Determine how well the forum met its objectives; if the data show the forum did not meet its objectives, determine the reasons and, as appropriate, modify the plan for an environmental leadership forum.

SAMPLE INVITATION LETTER

# *Environmental Leadership Forum*

*City of Olathe, Kansas; Kansas City, Missouri, Water Services Department*

*American Water Works Association Research Foundation*

Nov 7, 2003

Dear NAME,

You are cordially invited to the American Water Works Association Research Foundation's Environmental Leadership Forum for the metropolitan Kansas City area on December 2, 2003.

The Kansas City region is one of only three metropolitan areas in North America selected by the American Water Works Association Research Foundation (AwwaRF) to participate in this effort. **The outcome of this forum will be used to shape national and regional strategies for environmental leadership.**

| | |
|---|---|
| **When:** | Tuesday, December 2, 9:30 a.m. – 3:00 p.m. (lunch will be provided) |
| **Where:** | Municipal Services Center, the first certified "green" public building in Kansas 1385 Robinson, Olathe, KS 66061 (call 913-971-9311 if you need directions) |
| **Hosts:** | City of Olathe and the City of Kansas City, Missouri |
| **Questions:** | Contact Bob Cicerone, ETC Institute's event coordinator at 913-829-1215 |
| **Cost:** | It is free. Lunch, drinks, and snacks will be provided. |

**You will be joining nearly 50 civic leaders from the Kansas City region to discuss the role that local governments and public utilities should have with respect to environmental leadership.** The results of this initiative will be published by AwwaRF next year. As part of the forum, those attending will be provided with the results of a recent survey and other research that shows how residents of the Kansas City are compare to other communities across the United States. **Steve Maxwell**, a national authority on trends in the water industry from Boulder, Colorado, will be the featured speaker during lunch. He will discuss applications for environmental leadership in the Kansas City region.

We encourage you to help us make Environmental Leadership a priority for our community by attending this important event. **Please confirm your attendance by e-mailing DEC2Forum@aol.com or by calling Bob Cicerone at 913-829-1215 by Nov 20.**

Sincerely,


| | | |
|---|---|---|
| Michael Copeland | Frank Pogge | Chris Tatham |
| Mayor, City of Olathe | Director, KCMO WSD | Project Manager |

**SAMPLE AGENDA**

# AwwaRF Environmental Leadership Forum Agenda

930-100 0      Registration

930-101 5      Welcome/Overview of the Leadership Forum

1015-1130     Review of Research Completed to Date

1130-1150     Lunch

1150-1230     Guest Speaker

1230-1245     Break

1245-200       Break-Out Sessions/Focus Groups
- Development of Environmental Leadership Strategies

200-215         Break

215-245         Break-Out Groups Report Findings to the Whole Group

245-300         Final Wrap-Up/Next Steps

300               Adjourn

**American Water Works Association
Research Foundation
Environmental Leadership Forum**

**Break-Out Session
Leader's Guide**

**Olathe/KCMO
December 2, 2003**

**by**

**ETC Institute**

# Olathe/KCMO
# Environmental Leadership Forum
# BREAK-OUT SESSION LEADER GUIDE

## *Overview – 5 minutes*
- Re-introduce yourself

- Ask each person to introduce themselves (name, position, organization, and why they came to the meeting)

- Review the agenda for the breakout session: (1) Brainstorm Possible Environmental Leadership Strategies, (2) Prioritize Strategies, (3) Develop Specific Actions for the Top Strategies

- Review Rules
    - Turn cell phone ringers off
    - Everyone contributes
    - There are no right or wrong ideas
    - No internal conversations, share all comments with the entire group
    - Be Brief and Brilliant

## *Initial Brainstorming – 20 minutes*
What strategies could the City of Olathe/KCMO pursue to strengthen its position as an environmental leader? And why do you feel that way?

- Give the participants a few moments to <u>write down</u> their ideas.

- Have each person share ONE idea with the other members of the group; ask the person why they think their idea is important.

- After each person explains their idea, re-state their suggestion in your words to show you understand what was said and record the response on flip chart.

- Try to get at least 10 – 12 unique suggestions, mount additional flip chart pages on the wall if needed.

- Once the group is finished with this exercise, label each unique response on the flip charts with a letter (e.g., label each idea A, B, C, D, etc.).

## *Prioritizing the Strategies – 15 minutes*
- Distribute blank sheets of paper to the group.

- Ask participants to select the three ideas they think are the best suggestions and <u>write the letters for their top three choices</u> on the paper you provided.

- Once the group is finished, ask each person to tell you which three items they selected and why. Record a mark on the flip charts next to each item that is mentioned.

- Identify the 3-5 items that were picked by the most participants. These will be included in the next part of the discussion.

## *Developing Specific Action Strategies – 40 minutes*

Ask one member of the group to help you record responses.

Once someone has agreed to help record responses, discuss the following for each of the strategies that were identified as your group's top 3-5 priorities:

- Identify why this strategy is important and when it would be appropriate for Olathe/KCMO to implement the strategy.

- Identify the organization(s) and/or person(s) who should be involved, including the person or organization that should lead strategy.

- Determine the specific actions that Olathe/KCMO and the other organizations that were mentioned should take to implement the strategy.

- Identify obstacles or barriers might reduce the success of this strategy and what could be done to avoid or minimize the effects of these barriers or obstacle.

Have the recorder from your group write the responses for each strategy on the special worksheets that were provided. You should have 3-5 completed sets of worksheets when your group is finished.

## *Closing – 10 minutes*
- Ask each participant to share one final thought or comment before you end the discussion.

- Identify one person to be the spokesperson for the group. This person will briefly present the 3-5 strategies that your group developed to the entire forum.

- Tell your group to return to the main meeting room.

- Thank participants for their contributions.

# APPENDIX D
# SURVEY RESULTS

**Q1. I'd like to begin by asking you if you think your water utility is an environmental leader? An "environmental leader" would be an organization that is actively taking steps to protect the environment and conserve environmental resources.**

|  | Sample | | | |
|---|---|---|---|---|
|  | National<br>N | Ft Lauderdale<br>F | San Diego<br>S | Metro Kansas City<br>M |
| Q1 Is water utility environmental leader | | | | |
| 1=Yes | 30.7% | 31.4% | 30.0% | 33.1% |
| 2=No | 13.2% | 11.4% | 17.2% | 7.3% |
| 3=Don't know | 56.1% | 57.3% | 52.8% | 59.5% |

**Q2. Using a 5-point scale where 5 is very important and 1 is not important at all, please indicate how important you think it is for your water utility to be doing the following types of environmental leadership activities.**

|  | Sample | | | |
| --- | --- | --- | --- | --- |
|  | National<br>N | Ft Lauderdale<br>F | San Diego<br>S | Metro<br>Kansas City<br>M |
| Q2a Helping fund initiatives that help | | | | |
| 1=Not important at | 3.4% | 2.7% | 6.4% | 1.5% |
| 2=2 | 2.6% | 2.2% | 4.2% | 3.2% |
| 3=3 | 11.6% | 7.0% | 15.7% | 11.6% |
| 4=4 | 20.8% | 18.4% | 21.6% | 22.3% |
| 5=Very important | 58.8% | 68.6% | 49.6% | 57.7% |
| 9=Don't know | 2.7% | 1.1% | 2.5% | 3.7% |
| Q2b Investing in environmentally friend | | | | |
| 1=Not important at | 1.9% | 1.1% | 2.2% | 1.2% |
| 2=2 | 2.1% | 0.5% | 2.9% | 2.1% |
| 3=3 | 10.6% | 8.1% | 11.8% | 11.7% |
| 4=4 | 22.0% | 15.1% | 22.9% | 25.1% |
| 5=Very important | 60.5% | 71.9% | 57.7% | 55.6% |
| 9=Don't know | 2.9% | 3.2% | 2.5% | 4.2% |

**Q2. Using a 5-point scale where 5 is very important and 1 is not important at all, please indicate how important you think it is for your water utility to be doing the following types of environmental leadership activities.**

|  | Sample | | | |
|---|---|---|---|---|
|  | National N | Ft Lauderdale F | San Diego S | Metro Kansas City M |

### Q2c Cooperating with other organizations

|  | | | | |
|---|---|---|---|---|
| 1=Not important at | 1.0% | 1.1% | 0.7% | 0.5% |
| 2=2 | 0.2% | 0.5% | 1.7% | 0.6% |
| 3=3 | 4.0% | 0.5% | 2.9% | 4.2% |
| 4=4 | 12.1% | 9.7% | 12.8% | 11.9% |
| 5=Very important | 81.8% | 87.0% | 81.3% | 81.2% |
| 9=Don't know | 0.9% | 1.1% | 0.5% | 1.5% |

### Q2d Cooperating w/other org to prevent

|  | | | | |
|---|---|---|---|---|
| 1=Not important at | 1.0% | 2.2% | 2.7% | 1.0% |
| 2=2 | 1.6% | 0.0% | 2.7% | 1.2% |
| 3=3 | 6.7% | 6.5% | 8.6% | 6.0% |
| 4=4 | 15.1% | 12.4% | 19.4% | 19.8% |
| 5=Very important | 73.7% | 77.3% | 64.6% | 70.1% |
| 9=Don't know | 1.9% | 1.6% | 2.0% | 1.9% |

**Q2. Using a 5-point scale where 5 is very important and 1 is not important at all, please indicate how important you think it is for your water utility to be doing the following types of environmental leadership activities.**

|  | Sample | | | |
| --- | --- | --- | --- | --- |
|  | National<br>N | Ft Lauderdale<br>F | San Diego<br>S | Metro Kansas City<br>M |

Q2e Managing all forms of water

| | | | | |
| --- | --- | --- | --- | --- |
| 1=Not important at | 5.9% | 3.8% | 6.4% | 4.7% |
| 2=2 | 4.2% | 1.6% | 3.2% | 4.4% |
| 3=3 | 11.9% | 9.7% | 14.7% | 14.4% |
| 4=4 | 20.6% | 17.3% | 20.9% | 20.8% |
| 5=Very important | 46.9% | 53.0% | 45.9% | 45.3% |
| 9=Don't know | 10.5% | 14.6% | 8.8% | 10.4% |

Q2f Cooperating with other organizations

| | | | | |
| --- | --- | --- | --- | --- |
| 1=Not important at | 2.7% | 1.6% | 6.1% | 2.5% |
| 2=2 | 4.2% | 2.2% | 6.6% | 4.9% |
| 3=3 | 12.6% | 13.0% | 19.4% | 13.6% |
| 4=4 | 22.1% | 20.5% | 19.2% | 27.2% |
| 5=Very important | 55.7% | 60.0% | 44.7% | 49.0% |
| 9=Don't know | 2.6% | 2.7% | 3.9% | 2.9% |

**Q2. Using a 5-point scale where 5 is very important and 1 is not important at all, please indicate how important you think it is for your water utility to be doing the following types of environmental leadership activities.**

|  | Sample | | | |
|---|---|---|---|---|
|  | National N | Ft Lauderdale F | San Diego S | Metro Kansas City M |

**Q2g Informing the public about ways to**

| | | | | |
|---|---|---|---|---|
| 1=Not important at | 0.7% | 0.5% | 1.2% | 0.9% |
| 2=2 | 2.1% | 0.5% | 1.7% | 1.0% |
| 3=3 | 6.7% | 4.9% | 8.1% | 7.5% |
| 4=4 | 21.0% | 14.1% | 23.6% | 19.0% |
| 5=Very important | 68.5% | 79.5% | 63.6% | 70.6% |
| 9=Don't know | 0.9% | 0.5% | 1.7% | 1.0% |

**Q2h Anticipating how future development**

| | | | | |
|---|---|---|---|---|
| 1=Not important at | 0.6% | 1.1% | 1.2% | 0.4% |
| 2=2 | 1.2% | 0.0% | 2.0% | 0.6% |
| 3=3 | 5.7% | 2.7% | 7.6% | 6.1% |
| 4=4 | 18.5% | 15.1% | 22.9% | 21.8% |
| 5=Very important | 72.2% | 78.9% | 65.1% | 69.8% |
| 9=Don't know | 1.7% | 2.2% | 1.2% | 1.4% |

**Q2. Using a 5-point scale where 5 is very important and 1 is not important at all, please indicate how important you think it is for your water utility to be doing the following types of environmental leadership activities.**

| | Sample | | | |
|---|---|---|---|---|
| | National N | Ft Lauderdale F | San Diego S | Metro Kansas City M |

**Q2i Encouraging people to help protect**

| | | | | |
|---|---|---|---|---|
| 1=Not important at | 0.5% | 1.1% | 1.5% | 0.5% |
| 2=2 | 1.5% | 0.0% | 1.7% | 1.0% |
| 3=3 | 4.9% | 3.2% | 7.4% | 4.5% |
| 4=4 | 15.4% | 12.4% | 18.7% | 17.0% |
| 5=Very important | 77.2% | 82.7% | 70.0% | 75.9% |
| 9=Don't know | 0.6% | 0.5% | 0.7% | 1.1% |

**Q2j Encouraging water users in the region**

| | | | | |
|---|---|---|---|---|
| 1=Not important at | 1.1% | 1.6% | 1.5% | 0.9% |
| 2=2 | 2.6% | 1.1% | 0.7% | 1.7% |
| 3=3 | 6.4% | 6.5% | 6.4% | 9.8% |
| 4=4 | 20.8% | 14.1% | 24.6% | 23.6% |
| 5=Very important | 68.3% | 75.7% | 66.1% | 62.8% |
| 9=Don't know | 0.7% | 1.1% | 0.7% | 1.1% |

**Q2. Using a 5-point scale where 5 is very important and 1 is not important at all, please indicate how important you think it is for your water utility to be doing the following types of environmental leadership activities.**

| | Sample | | | |
|---|---|---|---|---|
| | National N | Ft Lauderdale F | San Diego S | Metro Kansas City M |

Q2k Acquiring land near lakes/streams

| | | | | |
|---|---|---|---|---|
| 1=Not important at | 3.5% | 3.2% | 2.5% | 2.0% |
| 2=2 | 4.7% | 1.6% | 4.9% | 3.2% |
| 3=3 | 10.2% | 7.6% | 14.7% | 11.8% |
| 4=4 | 16.5% | 17.8% | 21.6% | 21.9% |
| 5=Very important | 61.2% | 68.1% | 52.8% | 57.7% |
| 9=Don't know | 3.9% | 1.6% | 3.4% | 3.4% |

**Q2. Using a 5-point scale where 5 is very important and 1 is not important at all, please indicate how important you think it is for your water utility to be doing the following types of environmental leadership activities. (excluding don't knows)**

|  | Sample | | | |
|---|---|---|---|---|
|  | National N | Ft Lauderdale F | San Diego S | Metro Kansas City M |

Q2a Helping fund initiatives that help

| | | | | |
|---|---|---|---|---|
| 1=Not important at | 3.5% | 2.7% | 6.5% | 1.6% |
| 2=2 | 2.7% | 2.2% | 4.3% | 3.4% |
| 3=3 | 11.9% | 7.1% | 16.1% | 12.0% |
| 4=4 | 21.4% | 18.6% | 22.2% | 23.1% |
| 5=Very important | 60.5% | 69.4% | 50.9% | 59.9% |

Q2b Investing in environmentally friend

| | | | | |
|---|---|---|---|---|
| 1=Not important at | 1.9% | 1.1% | 2.3% | 1.3% |
| 2=2 | 2.2% | 0.6% | 3.0% | 2.2% |
| 3=3 | 10.9% | 8.4% | 12.1% | 12.2% |
| 4=4 | 22.6% | 15.6% | 23.4% | 26.2% |
| 5=Very important | 62.3% | 74.3% | 59.2% | 58.1% |

**Q2. Using a 5-point scale where 5 is very important and 1 is not important at all, please indicate how important you think it is for your water utility to be doing the following types of environmental leadership activities. (excluding don't knows)**

|  | Sample | | | |
|---|---|---|---|---|
|  | National N | Ft Lauderdale F | San Diego S | Metro Kansas City M |

Q2c Cooperating with other organizations

| | | | | |
|---|---|---|---|---|
| 1=Not important at | 1.0% | 1.1% | 0.7% | 0.5% |
| 2=2 | 0.3% | 0.5% | 1.7% | 0.6% |
| 3=3 | 4.0% | 0.5% | 3.0% | 4.3% |
| 4=4 | 12.2% | 9.8% | 12.8% | 12.1% |
| 5=Very important | 82.5% | 88.0% | 81.7% | 82.4% |

Q2d Cooperating w/other org to prevent

| | | | | |
|---|---|---|---|---|
| 1=Not important at | 1.0% | 2.2% | 2.8% | 1.0% |
| 2=2 | 1.7% | 0.0% | 2.8% | 1.3% |
| 3=3 | 6.9% | 6.6% | 8.8% | 6.1% |
| 4=4 | 15.4% | 12.6% | 19.8% | 20.2% |
| 5=Very important | 75.1% | 78.6% | 65.9% | 71.5% |

**Q2. Using a 5-point scale where 5 is very important and 1 is not important at all, please indicate how important you think it is for your water utility to be doing the following types of environmental leadership activities. (excluding don't knows)**

|  | Sample | | | |
|---|---|---|---|---|
|  | National<br>N | Ft Lauderdale<br>F | San Diego<br>S | Metro<br>Kansas City<br>M |

Q2e Managing all forms of water

| | | | | |
|---|---|---|---|---|
| 1=Not important at | 6.6% | 4.4% | 7.0% | 5.3% |
| 2=2 | 4.7% | 1.9% | 3.5% | 4.9% |
| 3=3 | 13.2% | 11.4% | 16.2% | 16.1% |
| 4=4 | 23.0% | 20.3% | 22.9% | 23.2% |
| 5=Very important | 52.4% | 62.0% | 50.4% | 50.6% |

Q2f Cooperating with other organizations

| | | | | |
|---|---|---|---|---|
| 1=Not important at | 2.8% | 1.7% | 6.4% | 2.6% |
| 2=2 | 4.4% | 2.2% | 6.9% | 5.0% |
| 3=3 | 12.9% | 13.3% | 20.2% | 14.0% |
| 4=4 | 22.7% | 21.1% | 19.9% | 28.0% |
| 5=Very important | 57.2% | 61.7% | 46.5% | 50.4% |

**Q2. Using a 5-point scale where 5 is very important and 1 is not important at all, please indicate how important you think it is for your water utility to be doing the following types of environmental leadership activities. (excluding don't knows)**

|  | Sample | | | |
|---|---|---|---|---|
|  | National N | Ft Lauderdale F | San Diego S | Metro Kansas City M |

Q2g Informing the public about ways to

| | | | | |
|---|---|---|---|---|
| 1=Not important at | 0.8% | 0.5% | 1.3% | 0.9% |
| 2=2 | 2.1% | 0.5% | 1.8% | 1.0% |
| 3=3 | 6.8% | 4.9% | 8.3% | 7.5% |
| 4=4 | 21.2% | 14.1% | 24.0% | 19.2% |
| 5=Very important | 69.1% | 79.9% | 64.8% | 71.4% |

Q2h Anticipating how future development

| | | | | |
|---|---|---|---|---|
| 1=Not important at | 0.6% | 1.1% | 1.2% | 0.4% |
| 2=2 | 1.3% | 0.0% | 2.0% | 0.6% |
| 3=3 | 5.8% | 2.8% | 7.7% | 6.2% |
| 4=4 | 18.8% | 15.5% | 23.1% | 22.1% |
| 5=Very important | 73.4% | 80.7% | 65.9% | 70.7% |

**Q2. Using a 5-point scale where 5 is very important and 1 is not important at all, please indicate how important you think it is for your water utility to be doing the following types of environmental leadership activities. (excluding don't knows)**

|  | Sample | | | |
| --- | --- | --- | --- | --- |
|  | National N | Ft Lauderdale F | San Diego S | Metro Kansas City M |

Q2i Encouraging people to help protect

| | | | | |
| --- | --- | --- | --- | --- |
| 1=Not important at | 0.5% | 1.1% | 1.5% | 0.5% |
| 2=2 | 1.5% | 0.0% | 1.7% | 1.0% |
| 3=3 | 4.9% | 3.3% | 7.4% | 4.5% |
| 4=4 | 15.5% | 12.5% | 18.8% | 17.2% |
| 5=Very important | 77.6% | 83.2% | 70.5% | 76.7% |

Q2j Encouraging water users in the region

| | | | | |
| --- | --- | --- | --- | --- |
| 1=Not important at | 1.1% | 1.6% | 1.5% | 0.9% |
| 2=2 | 2.6% | 1.1% | 0.7% | 1.8% |
| 3=3 | 6.4% | 6.6% | 6.4% | 9.9% |
| 4=4 | 21.0% | 14.2% | 24.8% | 23.9% |
| 5=Very important | 68.8% | 76.5% | 66.6% | 63.5% |

**Q2. Using a 5-point scale where 5 is very important and 1 is not important at all, please indicate how important you think it is for your water utility to be doing the following types of environmental leadership activities. (excluding don't knows)**

|  | Sample | | | |
|---|---|---|---|---|
|  | National<br>N | Ft Lauderdale<br>F | San Diego<br>S | Metro<br>Kansas City<br>M |

Q2k Acquiring land near lakes/streams

| | | | | |
|---|---|---|---|---|
| 1=Not important at | 3.6% | 3.3% | 2.5% | 2.1% |
| 2=2 | 4.9% | 1.6% | 5.1% | 3.3% |
| 3=3 | 10.6% | 7.7% | 15.3% | 12.2% |
| 4=4 | 17.1% | 18.1% | 22.4% | 22.7% |
| 5=Very important | 63.6% | 69.2% | 54.7% | 59.7% |

## Q3. Which THREE of the actions from the list I just read do you think are most important for your water utility to do? (all three selections)

| | Sample | | | |
|---|---|---|---|---|
| | National<br>N | Ft Lauderdale<br>F | San Diego<br>S | Metro Kansas City<br>M |
| **Q3 Sum of top 3 choices** | | | | |
| A=Help fund | 18.6% | 10.8% | 15.2% | 22.3% |
| B=Invest in | 24.2% | 30.3% | 25.1% | 21.9% |
| C=Cooperate w/ | 45.6% | 41.6% | 49.4% | 48.0% |
| D=Cooperate w/ | 30.6% | 20.5% | 20.9% | 26.6% |
| E=Manage all | 12.9% | 8.6% | 17.0% | 13.9% |
| F=Cooperate w/ | 17.9% | 8.1% | 12.3% | 13.3% |
| G=Inform public | 24.8% | 30.3% | 25.1% | 28.4% |
| H=Anticipate how | 21.0% | 29.2% | 25.6% | 25.7% |
| I=Encourage people | 28.7% | 38.9% | 31.0% | 27.2% |
| J=Encourage water | 20.5% | 27.6% | 26.3% | 18.3% |
| K=Acquire land | 21.7% | 25.9% | 22.9% | 21.4% |
| Z=None chosen | 10.4% | 8.6% | 6.9% | 9.2% |

**Q4. Using a 5-point scale where 5 is "Very Important" and 1 is "Not Important at all," please indicate how important you think it is for your water utility to do the following:**

|  | Sample | | | |
|---|---|---|---|---|
|  | National N | Ft Lauderdale F | San Diego S | Metro Kansas City M |

**Q4a Provide safe drinking water**

|  | | | | |
|---|---|---|---|---|
| 1=Not important at | 0.6% | 0.0% | 0.0% | 0.1% |
| 2=2 | 0.1% | 0.0% | 0.0% | 0.0% |
| 3=3 | 1.1% | 0.5% | 0.7% | 0.5% |
| 4=4 | 3.2% | 2.2% | 4.2% | 3.6% |
| 5=Very important | 94.9% | 97.3% | 95.1% | 95.3% |
| 9=Don't know | 0.0% | 0.0% | 0.0% | 0.5% |

**Q4b Provide good tasting drinking water**

|  | | | | |
|---|---|---|---|---|
| 1=Not important at | 1.0% | 1.6% | 1.7% | 0.2% |
| 2=2 | 1.0% | 0.0% | 1.0% | 0.2% |
| 3=3 | 6.5% | 5.4% | 12.0% | 4.5% |
| 4=4 | 19.0% | 7.0% | 24.8% | 17.7% |
| 5=Very important | 72.2% | 83.8% | 59.7% | 76.7% |
| 9=Don't know | 0.4% | 2.2% | 0.7% | 0.6% |

**Q4. Using a 5-point scale where 5 is "Very Important" and 1 is "Not Important at all," please indicate how important you think it is for your water utility to do the following:**

|  | Sample | | | |
|---|---|---|---|---|
|  | National<br>N | Ft Lauderdale<br>F | San Diego<br>S | Metro Kansas City<br>M |
| **Q4c Provide inexpensive drinking water** | | | | |
| 1=Not important at | 1.5% | 2.2% | 1.2% | 1.4% |
| 2=2 | 1.4% | 1.6% | 1.5% | 1.2% |
| 3=3 | 10.4% | 5.9% | 17.0% | 10.2% |
| 4=4 | 21.1% | 14.1% | 22.4% | 17.3% |
| 5=Very important | 65.4% | 74.6% | 57.5% | 69.0% |
| 9=Don't know | 0.2% | 1.6% | 0.5% | 0.9% |
| **Q4d Provide adequate water pressure to** | | | | |
| 1=Not important at | 0.6% | 1.1% | 0.0% | 0.1% |
| 2=2 | 1.1% | 1.1% | 0.7% | 0.5% |
| 3=3 | 6.2% | 3.2% | 10.1% | 3.9% |
| 4=4 | 19.6% | 14.1% | 25.6% | 21.9% |
| 5=Very important | 72.2% | 80.0% | 63.6% | 73.0% |
| 9=Don't know | 0.2% | 0.5% | 0.0% | 0.6% |

**Q4. Using a 5-point scale where 5 is "Very Important" and 1 is "Not Important at all," please indicate how important you think it is for your water utility to do the following:**

|  | Sample | | | |
|---|---|---|---|---|
|  | National<br>N | Ft Lauderdale<br>F | San Diego<br>S | Metro Kansas City<br>M |
| **Q4e Repair broken water mains quickly** | | | | |
| 1=Not important at | 0.5% | 0.0% | 0.2% | 0.0% |
| 2=2 | 0.2% | 0.0% | 0.0% | 0.1% |
| 3=3 | 2.9% | 1.6% | 3.2% | 2.0% |
| 4=4 | 10.2% | 4.3% | 11.5% | 10.2% |
| 5=Very important | 85.5% | 93.5% | 84.5% | 87.3% |
| 9=Don't know | 0.6% | 0.5% | 0.5% | 0.4% |
| **Q4f Provide accurate water bills** | | | | |
| 1=Not important at | 0.7% | 0.0% | 0.0% | 0.2% |
| 2=2 | 0.5% | 0.5% | 0.7% | 0.2% |
| 3=3 | 2.5% | 0.0% | 5.2% | 2.5% |
| 4=4 | 9.4% | 5.4% | 13.5% | 9.6% |
| 5=Very important | 86.1% | 93.0% | 79.6% | 86.8% |
| 9=Don't know | 0.7% | 1.1% | 1.0% | 0.6% |

**Q4. Using a 5-point scale where 5 is "Very Important" and 1 is "Not Important at all," please indicate how important you think it is for your water utility to do the following:**

|  | Sample | | | |
|---|---|---|---|---|
|  | National<br>N | Ft Lauderdale<br>F | San Diego<br>S | Metro Kansas City<br>M |
| **Q4g Provide residents with info about** | | | | |
| 1=Not important at | 1.4% | 0.0% | 0.7% | 0.4% |
| 2=2 | 2.6% | 0.0% | 3.7% | 1.0% |
| 3=3 | 9.7% | 8.6% | 18.2% | 10.2% |
| 4=4 | 21.5% | 22.2% | 23.6% | 24.8% |
| 5=Very important | 64.7% | 69.2% | 53.1% | 63.3% |
| 9=Don't know | 0.1% | 0.0% | 0.7% | 0.4% |
| **Q4h Protect the environment** | | | | |
| 1=Not important at | 1.2% | 0.5% | 0.5% | 0.4% |
| 2=2 | 0.9% | 0.5% | 2.5% | 0.6% |
| 3=3 | 4.2% | 3.2% | 7.6% | 4.0% |
| 4=4 | 11.4% | 10.3% | 17.9% | 15.0% |
| 5=Very important | 81.0% | 85.4% | 71.0% | 79.2% |
| 9=Don't know | 1.2% | 0.0% | 0.5% | 0.7% |

**Q4. Using a 5-point scale where 5 is "Very Important" and 1 is "Not Important at all," please indicate how important you think it is for your water utility to do the following: (excluding don't knows)**

|  | Sample | | | |
|---|---|---|---|---|
|  | National<br>N | Ft Lauderdale<br>F | San Diego<br>S | Metro Kansas City<br>M |
| **Q4a Provide safe drinking water** | | | | |
| 1=Not important at | 0.6% | 0.0% | 0.0% | 0.1% |
| 2=2 | 0.1% | 0.0% | 0.0% | 0.0% |
| 3=3 | 1.1% | 0.5% | 0.7% | 0.5% |
| 4=4 | 3.2% | 2.2% | 4.2% | 3.6% |
| 5=Very important | 94.9% | 97.3% | 95.1% | 95.8% |
| **Q4b Provide good tasting drinking water** | | | | |
| 1=Not important at | 1.0% | 1.7% | 1.7% | 0.3% |
| 2=2 | 1.0% | 0.0% | 1.0% | 0.3% |
| 3=3 | 6.5% | 5.5% | 12.1% | 4.5% |
| 4=4 | 19.0% | 7.2% | 25.0% | 17.8% |
| 5=Very important | 72.4% | 85.6% | 60.1% | 77.2% |

**Q4. Using a 5-point scale where 5 is "Very Important" and 1 is "Not Important at all," please indicate how important you think it is for your water utility to do the following: (excluding don't knows)**

|  | Sample | | | |
|---|---|---|---|---|
|  | National N | Ft Lauderdale F | San Diego S | Metro Kansas City M |

Q4c Provide inexpensive drinking water

| | | | | |
|---|---|---|---|---|
| 1=Not important at | 1.5% | 2.2% | 1.2% | 1.4% |
| 2=2 | 1.4% | 1.6% | 1.5% | 1.3% |
| 3=3 | 10.4% | 6.0% | 17.0% | 10.3% |
| 4=4 | 21.2% | 14.3% | 22.5% | 17.4% |
| 5=Very important | 65.6% | 75.8% | 57.8% | 69.6% |

Q4d Provide adequate water pressure to

| | | | | |
|---|---|---|---|---|
| 1=Not important at | 0.6% | 1.1% | 0.0% | 0.1% |
| 2=2 | 1.1% | 1.1% | 0.7% | 0.5% |
| 3=3 | 6.3% | 3.3% | 10.1% | 3.9% |
| 4=4 | 19.6% | 14.1% | 25.6% | 22.0% |
| 5=Very important | 72.3% | 80.4% | 63.6% | 73.5% |

**Q4. Using a 5-point scale where 5 is "Very Important" and 1 is "Not Important at all," please indicate how important you think it is for your water utility to do the following: (excluding don't knows)**

|  | Sample | | | |
| --- | --- | --- | --- | --- |
|  | National N | Ft Lauderdale F | San Diego S | Metro Kansas City M |

**Q4e Repair broken water mains quickly**

| | | | | |
| --- | --- | --- | --- | --- |
| 1=Not important at | 0.5% | 0.0% | 0.2% | 0.0% |
| 2=2 | 0.3% | 0.0% | 0.0% | 0.1% |
| 3=3 | 2.9% | 1.6% | 3.2% | 2.0% |
| 4=4 | 10.3% | 4.3% | 11.6% | 10.2% |
| 5=Very important | 86.1% | 94.0% | 84.9% | 87.6% |

**Q4f Provide accurate water bills**

| | | | | |
| --- | --- | --- | --- | --- |
| 1=Not important at | 0.8% | 0.0% | 0.0% | 0.3% |
| 2=2 | 0.5% | 0.5% | 0.7% | 0.3% |
| 3=3 | 2.5% | 0.0% | 5.2% | 2.5% |
| 4=4 | 9.4% | 5.5% | 13.6% | 9.6% |
| 5=Very important | 86.8% | 94.0% | 80.4% | 87.4% |

**Q4. Using a 5-point scale where 5 is "Very Important" and 1 is "Not Important at all," please indicate how important you think it is for your water utility to do the following: (excluding don't knows)**

|  | Sample | | | |
| --- | --- | --- | --- | --- |
|  | National N | Ft Lauderdale F | San Diego S | Metro Kansas City M |

Q4g Provide residents with info about

| | | | | |
| --- | --- | --- | --- | --- |
| 1=Not important at | 1.4% | 0.0% | 0.7% | 0.4% |
| 2=2 | 2.6% | 0.0% | 3.7% | 1.0% |
| 3=3 | 9.8% | 8.6% | 18.3% | 10.2% |
| 4=4 | 21.5% | 22.2% | 23.8% | 24.8% |
| 5=Very important | 64.8% | 69.2% | 53.5% | 63.5% |

Q4h Protect the environment

| | | | | |
| --- | --- | --- | --- | --- |
| 1=Not important at | 1.3% | 0.5% | 0.5% | 0.4% |
| 2=2 | 0.9% | 0.5% | 2.5% | 0.6% |
| 3=3 | 4.3% | 3.2% | 7.7% | 4.0% |
| 4=4 | 11.5% | 10.3% | 18.0% | 15.2% |
| 5=Very important | 82.0% | 85.4% | 71.4% | 79.8% |

**Q5. Which THREE of the items I just read do you think are most important for your water utility to do? (all three selections)**

|  | Sample | | | |
|---|---|---|---|---|
|  | National<br>N | Ft Lauderdale<br>F | San Diego<br>S | Metro Kansas City<br>M |

Q5 Sum of Top 3 Choices

| | | | | |
|---|---|---|---|---|
| A=Safe drinking | 87.8% | 92.4% | 91.2% | 87.2% |
| B=Good tasting | 41.8% | 44.9% | 38.3% | 40.9% |
| C=Inexpensive | 27.1% | 27.0% | 30.2% | 28.5% |
| D=Water pressure | 16.4% | 21.6% | 19.7% | 21.9% |
| E=Repair broken | 25.5% | 28.6% | 35.4% | 36.7% |
| F=Accurate water | 26.1% | 20.5% | 28.0% | 26.7% |
| G=Info about | 13.1% | 10.8% | 11.8% | 11.2% |
| H=Protect the | 41.1% | 41.6% | 37.1% | 31.5% |
| Z=None chosen | 6.7% | 3.2% | 2.0% | 3.4% |

**Q6. Now I'm going to ask about your satisfaction with your water utility. Using a 5-point scale where 1 means "Very Dissatisfied" and 5 means "Very Satisfied," please rate your overall satisfaction with each of the following.**

|  | Sample | | | |
|---|---|---|---|---|
|  | National<br>N | Ft Lauderdale<br>F | San Diego<br>S | Metro Kansas City<br>M |
| Q6a How well water utility cooperates with | | | | |
| 1=Very dissatis | 2.9% | 2.2% | 1.2% | 1.5% |
| 2=Dissatisfied | 4.0% | 3.8% | 3.4% | 2.0% |
| 3=Neutral | 15.0% | 9.2% | 17.2% | 19.2% |
| 4=Satisfied | 21.0% | 16.2% | 18.9% | 16.5% |
| 5=Very satisfied | 17.4% | 15.1% | 14.5% | 18.3% |
| 9=Don't know | 39.8% | 53.5% | 44.7% | 42.5% |

**Q6. Now I'm going to ask about your satisfaction with your water utility. Using a 5-point scale where 1 means "Very Dissatisfied" and 5 means "Very Satisfied," please rate your overall satisfaction with each of the following.**

|  | Sample | | | |
|---|---|---|---|---|
|  | National<br>N | Ft Lauderdale<br>F | San Diego<br>S | Metro<br>Kansas City<br>M |
| **Q6b How often your water utility ask residents** | | | | |
| 1=Very dissatis | 19.5% | 14.1% | 18.4% | 19.0% |
| 2=Dissatisfied | 16.9% | 16.8% | 15.5% | 16.4% |
| 3=Neutral | 16.7% | 16.8% | 20.4% | 23.3% |
| 4=Satisfied | 13.5% | 7.0% | 15.2% | 11.6% |
| 5=Very satisfied | 10.4% | 9.7% | 7.1% | 9.3% |
| 9=Don't know | 23.1% | 35.7% | 23.3% | 20.4% |
| **Q6c How proactive your utility is being** | | | | |
| 1=Very dissatis | 5.6% | 3.2% | 3.4% | 3.2% |
| 2=Dissatisfied | 6.0% | 2.7% | 8.6% | 5.3% |
| 3=Neutral | 19.6% | 15.1% | 19.9% | 21.8% |
| 4=Satisfied | 19.7% | 15.7% | 20.9% | 16.4% |
| 5=Very satisfied | 17.6% | 14.1% | 16.0% | 18.7% |
| 9=Don't know | 31.5% | 49.2% | 31.2% | 34.6% |

**Q6. Now I'm going to ask about your satisfaction with your water utility. Using a 5-point scale where 1 means "Very Dissatisfied" and 5 means "Very Satisfied," please rate your overall satisfaction with each of the following.**

|  | Sample | | | |
|---|---|---|---|---|
|  | National<br>N | Ft Lauderdale<br>F | San Diego<br>S | Metro Kansas City<br>M |
| **Q6d How well water utility educate people** | | | | |
| 1=Very dissatis | 12.6% | 11.4% | 11.3% | 11.3% |
| 2=Dissatisfied | 14.1% | 13.5% | 16.0% | 16.4% |
| 3=Neutral | 24.3% | 17.3% | 23.8% | 23.6% |
| 4=Satisfied | 17.1% | 15.7% | 18.2% | 17.0% |
| 5=Very satisfied | 12.9% | 14.1% | 14.0% | 13.4% |
| 9=Don't know | 19.0% | 28.1% | 16.7% | 18.2% |
| **Q6e Overall how satisfied w/water supplier** | | | | |
| 1=Very dissatis | 5.4% | 5.9% | 2.2% | 3.0% |
| 2=Dissatisfied | 6.2% | 5.9% | 5.7% | 4.6% |
| 3=Neutral | 21.5% | 23.8% | 26.5% | 22.8% |
| 4=Satisfied | 36.5% | 33.0% | 40.0% | 40.2% |
| 5=Very satisfied | 26.3% | 25.9% | 21.4% | 27.4% |
| 9=Don't know | 4.1% | 5.4% | 4.2% | 2.1% |

**Q6. Now I'm going to ask about your satisfaction with your water utility. Using a 5-point scale where 1 means "Very Dissatisfied" and 5 means "Very Satisfied," please rate your overall satisfaction with each of the following. (excluding don't knows)**

|  | Sample | | | |
|---|---|---|---|---|
|  | National<br>N | Ft Lauderdale<br>F | San Diego<br>S | Metro Kansas City<br>M |

Q6a How well water utility cooperates with

| | National | Ft Lauderdale | San Diego | Metro Kansas City |
|---|---|---|---|---|
| 1=Very dissatis | 4.8% | 4.7% | 2.2% | 2.6% |
| 2=Dissatisfied | 6.6% | 8.1% | 6.2% | 3.5% |
| 3=Neutral | 24.9% | 19.8% | 31.1% | 33.3% |
| 4=Satisfied | 34.9% | 34.9% | 34.2% | 28.8% |
| 5=Very satisfied | 28.8% | 32.6% | 26.2% | 31.8% |

Q6b How often your water utility ask residents

| | National | Ft Lauderdale | San Diego | Metro Kansas City |
|---|---|---|---|---|
| 1=Very dissatis | 25.3% | 21.8% | 24.0% | 23.9% |
| 2=Dissatisfied | 21.9% | 26.1% | 20.2% | 20.6% |
| 3=Neutral | 21.8% | 26.1% | 26.6% | 29.2% |
| 4=Satisfied | 17.5% | 10.9% | 19.9% | 14.5% |
| 5=Very satisfied | 13.5% | 15.1% | 9.3% | 11.7% |

**Q6. Now I'm going to ask about your satisfaction with your water utility. Using a 5-point scale where 1 means "Very Dissatisfied" and 5 means "Very Satisfied," please rate your overall satisfaction with each of the following. (excluding don't knows)**

|  | Sample | | | |
|---|---|---|---|---|
|  | National<br>N | Ft Lauderdale<br>F | San Diego<br>S | Metro<br>Kansas City<br>M |

Q6c How proactive your utility is being

| | | | | |
|---|---|---|---|---|
| 1=Very dissatis | 8.2% | 6.4% | 5.0% | 4.9% |
| 2=Dissatisfied | 8.7% | 5.3% | 12.5% | 8.2% |
| 3=Neutral | 28.6% | 29.8% | 28.9% | 33.3% |
| 4=Satisfied | 28.8% | 30.9% | 30.4% | 25.1% |
| 5=Very satisfied | 25.7% | 27.7% | 23.2% | 28.5% |
| | | | | |
| 1=Very dissatis | 15.6% | 15.8% | 13.6% | 13.8% |
| 2=Dissatisfied | 17.4% | 18.8% | 19.2% | 20.1% |
| 3=Neutral | 30.0% | 24.1% | 28.6% | 28.9% |
| 4=Satisfied | 21.1% | 21.8% | 21.8% | 20.8% |
| 5=Very satisfied | 15.9% | 19.5% | 16.8% | 16.4% |

**Q6. Now I'm going to ask about your satisfaction with your water utility. Using a 5-point scale where 1 means "Very Dissatisfied" and 5 means "Very Satisfied," please rate your overall satisfaction with each of the following. (excluding don't knows)**

|  | Sample | | | |
|---|---|---|---|---|
|  | National N | Ft Lauderdale F | San Diego S | Metro Kansas City M |

Q6e Overall how satisfied w/water supplier

| | | | | |
|---|---|---|---|---|
| 1=Very dissatis | 5.6% | 6.3% | 2.3% | 3.0% |
| 2=Dissatisfied | 6.5% | 6.3% | 5.9% | 4.7% |
| 3=Neutral | 22.4% | 25.1% | 27.7% | 23.3% |
| 4=Satisfied | 38.0% | 34.9% | 41.8% | 41.0% |
| 5=Very satisfied | 27.5% | 27.4% | 22.3% | 28.0% |

**Q7. If your water supplier were recognized as a leader in protecting and preserving sources of drinking water would you be willing to pay a little more for your drinking water?**

|  | Sample | | | |
|---|---|---|---|---|
|  | National<br>N | Ft Lauderdale<br>F | San Diego<br>S | Metro<br>Kansas City<br>M |
| Q7 Willing to pay more if recognized | | | | |
| 1=Yes | 46.2% | 45.9% | 52.8% | 42.8% |
| 2=No | 33.2% | 36.8% | 33.7% | 35.9% |
| 9=Don't know | 20.6% | 17.3% | 13.5% | 21.3% |

**Q7a. IF YES, how much extra, in dollars, would you be willing to pay per month to cover the costs of these activities?**

| Q7a How much more per month sample | Mean |
|---|---|
| N=National | $7.53 |
| F=Ft Lauderdale | $5.55 |
| S=San Diego | $9.71 |
| M=Metro Kansas City | $5.84 |

**Q7a. IF YES, how much extra, in dollars, would you be willing to pay per month to cover the costs of these activities?**

|  | Sample | | | |
|---|---|---|---|---|
|  | National<br>N | Ft Lauderdale<br>F | San Diego<br>S | Metro<br>Kansas City<br>M |
| Q7a How much more per month | | | | |
| 1=$1 or less | 7.6% | 8.1% | 4.4% | 12.9% |
| 2=$1.01-$2 | 17.1% | 23.0% | 8.3% | 18.6% |
| 3=$2.01-$3 | 7.3% | 13.5% | 10.0% | 8.7% |
| 4=$3.01-$4 | 1.9% | 2.7% | 2.8% | 1.9% |
| 5=$4.01-$5 | 30.1% | 24.3% | 28.3% | 32.2% |
| 10=$5.01-$10 | 24.4% | 23.0% | 25.6% | 18.2% |
| 20=$10.01-$20 | 7.6% | 5.4% | 14.4% | 6.4% |
| 21=$20.01+ | 4.1% | 0.0% | 6.1% | 1.1% |

**Q8. Using a 5-point scale where 1 means "Not Supportive At All" and 5 means "Very Supportive," please indicate how supportive you would be of having your water utility do each of the following:**

|  | Sample | | | |
|---|---|---|---|---|
|  | National<br>N | Ft Lauderdale<br>F | San Diego<br>S | Metro Kansas City<br>M |
| **Q8a Investing in water treatment eqpt** | | | | |
| 1=Not support at | 2.2% | 1.6% | 3.7% | 1.1% |
| 2=Not supportive | 2.1% | 0.5% | 3.7% | 1.9% |
| 3=Neutral | 11.2% | 8.6% | 13.5% | 14.9% |
| 4=Supportive | 27.8% | 20.5% | 28.3% | 30.2% |
| 5=Very supportive | 53.9% | 64.3% | 47.9% | 48.6% |
| 9=Don't know | 2.6% | 4.3% | 2.9% | 3.2% |
| **Q8b Working with other governmental org** | | | | |
| 1=Not support at | 3.9% | 2.7% | 5.2% | 1.9% |
| 2=Not supportive | 3.4% | 3.2% | 3.9% | 3.2% |
| 3=Neutral | 16.1% | 12.4% | 17.9% | 18.5% |
| 4=Supportive | 27.1% | 25.9% | 27.8% | 31.8% |
| 5=Very supportive | 46.6% | 51.4% | 40.0% | 40.0% |
| 9=Don't know | 3.0% | 4.3% | 5.2% | 4.5% |

**Q8. Using a 5-point scale where 1 means "Not Supportive At All" and 5 means "Very Supportive," please indicate how supportive you would be of having your water utility do each of the following:**

|  | Sample | | | |
|---|---|---|---|---|
|  | National N | Ft Lauderdale F | San Diego S | Metro Kansas City M |
| **Q8c Spending money to educate people** | | | | |
| 1=Not support at | 4.7% | 6.5% | 4.7% | 3.0% |
| 2=Not supportive | 5.6% | 2.2% | 5.7% | 5.4% |
| 3=Neutral | 22.8% | 24.9% | 23.1% | 28.6% |
| 4=Supportive | 26.6% | 21.6% | 29.7% | 29.5% |
| 5=Very supportive | 37.2% | 43.2% | 34.4% | 29.9% |
| 9=Don't know | 3.0% | 1.6% | 2.5% | 3.6% |
| **Q8d Educate people about ways to protect** | | | | |
| 1=Not support at | 3.7% | 5.9% | 4.9% | 2.7% |
| 2=Not supportive | 5.0% | 3.2% | 3.9% | 5.0% |
| 3=Neutral | 16.7% | 16.2% | 20.1% | 20.6% |
| 4=Supportive | 27.2% | 23.8% | 26.5% | 29.9% |
| 5=Very supportive | 44.8% | 49.2% | 42.3% | 38.4% |
| 9=Don't know | 2.5% | 1.6% | 2.2% | 3.4% |

**Q8. Using a 5-point scale where 1 means "Not Supportive At All" and 5 means "Very Supportive," please indicate how supportive you would be of having your water utility do each of the following:**

| | Sample | | | |
|---|---|---|---|---|
| | National N | Ft Lauderdale F | San Diego S | Metro Kansas City M |

**Q8e Educate people in your community**

| | | | | |
|---|---|---|---|---|
| 1=Not support at | 4.5% | 5.9% | 7.6% | 3.7% |
| 2=Not supportive | 5.7% | 3.8% | 8.1% | 6.2% |
| 3=Neutral | 19.5% | 16.2% | 19.9% | 22.1% |
| 4=Supportive | 23.5% | 25.9% | 25.8% | 29.7% |
| 5=Very supportive | 43.8% | 46.5% | 36.4% | 34.2% |
| 9=Don't know | 3.0% | 1.6% | 2.2% | 4.0% |

**Q8f Encouraging local governments to**

| | | | | |
|---|---|---|---|---|
| 1=Not support at | 4.2% | 3.8% | 3.2% | 4.2% |
| 2=Not supportive | 4.0% | 4.3% | 4.2% | 4.2% |
| 3=Neutral | 14.7% | 12.4% | 16.7% | 19.0% |
| 4=Supportive | 27.1% | 21.6% | 24.6% | 27.7% |
| 5=Very supportive | 47.9% | 53.0% | 49.6% | 40.7% |
| 9=Don't know | 2.0% | 4.9% | 1.7% | 4.1% |

**Q8. Using a 5-point scale where 1 means "Not Supportive At All" and 5 means "Very Supportive," please indicate how supportive you would be of having your water utility do each of the following:**

|  | Sample | | | |
|---|---|---|---|---|
|  | National<br>N | Ft Lauderdale<br>F | San Diego<br>S | Metro Kansas City<br>M |
| **Q8g Restricting times and days that** | | | | |
| 1=Not support at | 9.5% | 7.0% | 10.3% | 10.1% |
| 2=Not supportive | 8.5% | 7.0% | 11.3% | 7.5% |
| 3=Neutral | 25.0% | 27.0% | 18.2% | 24.3% |
| 4=Supportive | 20.7% | 17.3% | 25.1% | 23.5% |
| 5=Very supportive | 33.8% | 38.4% | 32.9% | 31.0% |
| 9=Don't know | 2.5% | 3.2% | 2.2% | 3.7% |

**Q8. Using a 5-point scale where 1 means "Not Supportive At All" and 5 means "Very Supportive," please indicate how supportive you would be of having your water utility do each of the following: (excluding don't knows)**

|  | Sample | | | |
|---|---|---|---|---|
|  | National N | Ft Lauderdale F | San Diego S | Metro Kansas City M |

**Q8a Investing in water treatment eqpt**

|  | | | | |
|---|---|---|---|---|
| 1=Not support at | 2.3% | 1.7% | 3.8% | 1.2% |
| 2=Not supportive | 2.2% | 0.6% | 3.8% | 1.9% |
| 3=Neutral | 11.5% | 9.0% | 13.9% | 15.4% |
| 4=Supportive | 28.6% | 21.5% | 29.1% | 31.2% |
| 5=Very supportive | 55.4% | 67.2% | 49.4% | 50.3% |

**Q8b Working with other governmental org**

|  | | | | |
|---|---|---|---|---|
| 1=Not support at | 4.0% | 2.8% | 5.4% | 2.0% |
| 2=Not supportive | 3.5% | 3.4% | 4.1% | 3.4% |
| 3=Neutral | 16.6% | 13.0% | 18.9% | 19.4% |
| 4=Supportive | 27.9% | 27.1% | 29.3% | 33.3% |
| 5=Very supportive | 48.0% | 53.7% | 42.2% | 41.9% |

**Q8. Using a 5-point scale where 1 means "Not Supportive At All" and 5 means "Very Supportive," please indicate how supportive you would be of having your water utility do each of the following: (excluding don't knows)**

| | Sample | | | |
|---|---|---|---|---|
| | National N | Ft Lauderdale F | San Diego S | Metro Kansas City M |

### Q8c Spending money to educate people

| | | | | |
|---|---|---|---|---|
| 1=Not support at | 4.9% | 6.6% | 4.8% | 3.1% |
| 2=Not supportive | 5.8% | 2.2% | 5.8% | 5.6% |
| 3=Neutral | 23.6% | 25.3% | 23.7% | 29.7% |
| 4=Supportive | 27.4% | 22.0% | 30.5% | 30.6% |
| 5=Very supportive | 38.4% | 44.0% | 35.3% | 31.0% |

### Q8d Educate people about ways to protect

| | | | | |
|---|---|---|---|---|
| 1=Not support at | 3.8% | 6.0% | 5.0% | 2.8% |
| 2=Not supportive | 5.1% | 3.3% | 4.0% | 5.1% |
| 3=Neutral | 17.2% | 16.5% | 20.6% | 21.4% |
| 4=Supportive | 27.9% | 24.2% | 27.1% | 30.9% |
| 5=Very supportive | 46.0% | 50.0% | 43.2% | 39.8% |

**Q8. Using a 5-point scale where 1 means "Not Supportive At All" and 5 means "Very Supportive," please indicate how supportive you would be of having your water utility do each of the following: (excluding don't knows)**

|  | Sample | | | |
|---|---|---|---|---|
|  | National N | Ft Lauderdale F | San Diego S | Metro Kansas City M |
| **Q8e Educate people in your community** | | | | |
| 1=Not support at | 4.6% | 6.0% | 7.8% | 3.9% |
| 2=Not supportive | 5.9% | 3.8% | 8.3% | 6.5% |
| 3=Neutral | 20.1% | 16.5% | 20.4% | 23.1% |
| 4=Supportive | 24.2% | 26.4% | 26.4% | 31.0% |
| 5=Very supportive | 45.2% | 47.3% | 37.2% | 35.6% |
| **Q8f Encouraging local governments to** | | | | |
| 1=Not support at | 4.3% | 4.0% | 3.3% | 4.4% |
| 2=Not supportive | 4.1% | 4.5% | 4.3% | 4.4% |
| 3=Neutral | 15.0% | 13.1% | 17.0% | 19.8% |
| 4=Supportive | 27.6% | 22.7% | 25.0% | 28.9% |
| 5=Very supportive | 48.9% | 55.7% | 50.5% | 42.4% |

**Q8. Using a 5-point scale where 1 means "Not Supportive At All" and 5 means "Very Supportive," please indicate how supportive you would be of having your water utility do each of the following: (excluding don't knows)**

|  | Sample | | | |
| --- | --- | --- | --- | --- |
|  | National N | Ft Lauderdale F | San Diego S | Metro Kansas City M |
| Q8g Restricting times and days that | | | | |
| 1=Not support at | 9.7% | 7.3% | 10.6% | 10.5% |
| 2=Not supportive | 8.7% | 7.3% | 11.6% | 7.8% |
| 3=Neutral | 25.6% | 27.9% | 18.6% | 25.2% |
| 4=Supportive | 21.3% | 17.9% | 25.6% | 24.4% |
| 5=Very supportive | 34.7% | 39.7% | 33.7% | 32.2% |

**Q9. How supportive would you be of giving your water utility the authority to restrict development and recreational activity near lakes and other water ways that are used as sources of drinking water?**

| | Sample | | | |
|---|---|---|---|---|
| | National N | Ft Lauderdale F | San Diego S | Metro Kansas City M |

Q9 Support of giving authority-restrict

| | | | | |
|---|---|---|---|---|
| 1=Very supportive | 32.5% | 39.5% | 32.7% | 32.6% |
| 2=Somewhat | 40.7% | 36.2% | 40.0% | 38.8% |
| 3=Neutral | 15.1% | 11.4% | 11.8% | 17.7% |
| 4=Not supportive | 10.1% | 13.0% | 14.7% | 10.1% |
| 9=Don't know | 1.6% | 0.0% | 0.7% | 0.9% |

**Q10. Overall, how well informed do you think you are about environmental issues?**

| | Sample | | | |
|---|---|---|---|---|
| | National N | Ft Lauderdale F | San Diego S | Metro Kansas City M |

Q10 How informed are you about environment

| | | | | |
|---|---|---|---|---|
| 1=Very informed | 17.0% | 21.6% | 26.3% | 12.6% |
| 2=Somewhat | 55.4% | 57.8% | 55.3% | 55.0% |
| 3=Neutral | 15.0% | 9.7% | 8.1% | 14.8% |
| 4=Not informed | 12.5% | 10.8% | 10.3% | 17.4% |
| 9=Don't know | 0.1% | 0.0% | 0.0% | 0.2% |

**Q11. Overall, how important do you think it is for your water utility to be an Environmental Leader?**

|  | Sample | | | |
|---|---|---|---|---|
|  | National N | Ft Lauderdale F | San Diego S | Metro Kansas City M |
| Q11 Utility to be environmental leader | | | | |
| 1=Very important | 63.5% | 67.6% | 59.2% | 57.7% |
| 2=Somewhat | 26.8% | 27.6% | 30.5% | 33.1% |
| 3=Not sure | 6.5% | 3.8% | 5.2% | 7.0% |
| 4=Not important | 3.1% | 1.1% | 5.2% | 2.2% |

**Q12. In your opinion, which of the following best describes the overall attitude of the community where you live toward the environment? Would you say your community is generally:**

|  | Sample | | | |
|---|---|---|---|---|
|  | National N | Ft Lauderdale F | San Diego S | Metro Kansas City M |
| Q12 Overall attitude of community toward | | | | |
| 1=Very willing to | 13.6% | 17.8% | 18.7% | 14.1% |
| 2=Somewhat willing | 69.4% | 64.3% | 62.2% | 65.0% |
| 3=Somewhat unwilling | 11.7% | 10.3% | 12.5% | 12.7% |
| 4=Very unwilling | 3.0% | 5.9% | 3.2% | 3.0% |
| 9=Don't know | 2.2% | 1.6% | 3.4% | 5.2% |

## Q13. Which of the following best describes the community where you live:

| | Sample | | | |
|---|---|---|---|---|
| | National N | Ft Lauderdale F | San Diego S | Metro Kansas City M |

Q13 Description of community where you

| | National N | Ft Lauderdale F | San Diego S | Metro Kansas City M |
|---|---|---|---|---|
| 1=Large urban city | 20.0% | 37.3% | 57.5% | 32.0% |
| 2=Suburb of major city | 32.6% | 42.2% | 35.6% | 52.2% |
| 3=City outside a metropolitan area | 25.3% | 18.4% | 5.7% | 15.2% |
| 4=Rural area | 22.1% | 2.2% | 1.2% | 0.6% |

## Q14. Do you own or rent your home?

| | Sample | | | |
|---|---|---|---|---|
| | National N | Ft Lauderdale F | San Diego S | Metro Kansas City M |

Q14 Do you Own or rent your home

| | National N | Ft Lauderdale F | San Diego S | Metro Kansas City M |
|---|---|---|---|---|
| 1=Own | 77.9% | 84.9% | 68.0% | 69.6% |
| 2=Rent | 22.1% | 15.1% | 32.0% | 30.4% |

**Q15. How would you describe your race/ethnicity (please check all that apply)?**

| Q15 Respondents Race/ethnicity | Sample | | | |
|---|---|---|---|---|
| | National N | Ft Lauderdale F | San Diego S | Metro Kansas City M |
| 1=Hispanic | 11.6% | 17.2% | 25.0% | 5.2% |
| 2=White | 71.4% | 60.1% | 61.5% | 67.4% |
| 3=Amer Indian/ Esk | 0.9% | 2.2% | 0.7% | 1.9% |
| 4=Blk/African Am | 14.0% | 17.8% | 4.4% | 21.4% |
| 5=Asian/Pac Island | 1.5% | 2.2% | 7.6% | 2.2% |
| 6=Other | 0.5% | 0.5% | 1.0% | 0.6% |
| 9=No answer | 0.5% | 0.5% | 0.5% | 1.9% |

**Q16. How many persons are in your household (counting yourself)?**

| total persons | Mean |
|---|---|
| sample | |
| N=National | 2.68 persons |
| F=Ft Lauderdale | 2.70 |
| S=San Diego | 2.77 |
| M=Metro Kansas City | 2.91 |

## Q17. What is your total annual household income?

|  | Sample | | | |
|---|---|---|---|---|
|  | National<br>N | Ft Lauderdale<br>F | San Diego<br>S | Metro Kansas City<br>M |
| Q17 Total annual household income | | | | |
| 1=Under $25,000 | 17.2% | 15.7% | 10.3% | 18.0% |
| 2=$25,000-$49,999 | 28.6% | 27.6% | 23.1% | 28.7% |
| 3=$50,000-$74,999 | 22.1% | 21.1% | 22.6% | 17.7% |
| 4=$75,000-$99,999 | 8.9% | 8.1% | 10.3% | 8.1% |
| 5=$100,000+ | 9.0% | 14.1% | 13.8% | 5.5% |
| 9=Not provided | 14.2% | 13.5% | 19.9% | 22.0% |

Date: _____  Phone: _____  Interviewer: _____

# AwwaRF Environmental Leadership Survey

## *NATIONAL SURVEY INTRO*
This is _____ calling for ETC Institute. ETC Institute is working with the American Water Works Association Research Foundation to gather input from residents in your community about the role that water utilities should have in environmental leadership. Do you have time to answer a few questions? The survey takes less than 10 minutes. [If Asked: The American Water Works Association is a non-profit association of water utilities from around the world; the organization is based in Denver, Colorado].

## *LOCAL INTRO*
This is _____ calling for the City of [Olathe, Kansas City, Fort Lauderdale, or San Diego]. The reason I am calling is that the City of [Olathe, Kansas City, Fort Lauderdale, or San Diego] is working with the American Water Works Association Research Foundation to gather input from residents about the role that water utilities should have in environmental leadership. Do you have time to answer a few questions? The survey takes less than 10 minutes. [If Asked: The American Water Works Association is a non-profit association of water utilities from around the world; the organization is based in Denver, Colorado].

1. I'd like to begin by asking you if you think your water utility is an environmental leader? An "environmental leader" would be an organization that is actively taking steps to protect the environment and conserve environmental resources.
   ___(1) Yes
   ___(2) No
   ___(3) Don't know

   1a. Why do you feel that way?

   _____

2. **Using a 5-point scale where 5 is very important and 1 is not important at all, please indicate how important you think it is for your water utility to be doing the following types of environmental leadership activities.**

|   | | Not Important At All | | | | Very Important | Don't Know |
|---|---|---|---|---|---|---|---|
| A) | Helping fund initiatives that help protect the environment in the area where you live. | 1 | 2 | 3 | 4 | 5 | 9 |
| B) | Investing in environmentally friendly equipment and technology | 1 | 2 | 3 | 4 | 5 | 9 |
| C) | Cooperating with other organizations to prevent YOUR source of drinking water from being polluted. | 1 | 2 | 3 | 4 | 5 | 9 |
| D) | Cooperating with other organizations to prevent ALL sources of drinking water from being polluted, including those that are NOT sources of drinking water for your community. | 1 | 2 | 3 | 4 | 5 | 9 |
| E) | Managing all forms of water, including drinking water, storm water, waste water, and sewage, in an integrated manner rather than separately. | 1 | 2 | 3 | 4 | 5 | 9 |
| F) | Cooperating with other organizations to protect air quality and other aspects of the environment that are not directly related to water quality in the region | 1 | 2 | 3 | 4 | 5 | 9 |
| G) | Informing the public about ways to protect the sources of drinking water for your community. | 1 | 2 | 3 | 4 | 5 | 9 |
| H) | Anticipating how future development may affect the supply of safe drinking water in the area where you live | 1 | 2 | 3 | 4 | 5 | 9 |
| I) | Encouraging people to help protect sources of drinking water from contamination | 1 | 2 | 3 | 4 | 5 | 9 |
| J) | Encouraging water users in the region to conserve water | 1 | 2 | 3 | 4 | 5 | 9 |
| K) | Acquiring land near lakes and streams to protect water sources in the area where you live from contamination | 1 | 2 | 3 | 4 | 5 | 9 |

3. **Which THREE of the actions from the list I just read (Question 2) do you think are most important for your water utility to do?** (Reread the ones the person rated as very important "5" and write in the corresponding letters in the space below)

        1st:____      2nd:____      3rd:____      none of these

## *PERCEIVED IMPORTANCE OF WATER UTILITY FUNCTIONS*

4. **Using a 5-point scale where 5 is "Very Important" and 1 is "Not Important at all," please indicate how important you think it is for your water utility to do the following:**

|   |   | Not Important At All | | | | Very Important | Don't Know |
|---|---|---|---|---|---|---|---|
| A) | Provide safe drinking water. | 1 | 2 | 3 | 4 | 5 | 9 |
| B) | Provide good tasting drinking water | 1 | 2 | 3 | 4 | 5 | 9 |
| C) | Provide inexpensive drinking water | 1 | 2 | 3 | 4 | 5 | 9 |
| D) | Provide adequate water pressure to homes. | 1 | 2 | 3 | 4 | 5 | 9 |
| E) | Repair broken water mains quickly. | 1 | 2 | 3 | 4 | 5 | 9 |
| F) | Provide accurate water bills. | 1 | 2 | 3 | 4 | 5 | 9 |
| G) | Provide residents with information about water related issues. | 1 | 2 | 3 | 4 | 5 | 9 |
| H) | Protect the environment. | 1 | 2 | 3 | 4 | 5 | 9 |

5. **Which THREE of the items I just read (Question 4) do you think are most important for your water utility to do?** (Reread the ones the person rated as very important "5" and write in the corresponding letters in the space below)

   ____   ____   ____   None of these
   1st    2nd    3rd

## *SATISFACTION WITH LOCAL WATER UTILITY'S ENVIRONMENTAL LEADERSHIP*

6. **Now I'm going to ask about your satisfaction with your water utility. Using a 5-point scale where 1 means "Very Dissatisfied" and 5 means "Very Satisfied," please rate your overall satisfaction with each of the following.**

|   |   | Very Dissatisfied | Dissatisfied | Neutral | Satisfied | Very Satisfied | Don't Know |
|---|---|---|---|---|---|---|---|
| A) | How well your water utility cooperates with other organizations to protect and preserve sources of drinking water. | 1 | 2 | 3 | 4 | 5 | 9 |
| B) | How often your water utility asks resident for their ideas to protect and preserve sources of drinking water | 1 | 2 | 3 | 4 | 5 | 9 |
| C) | How proactive your utility is being to protect drinking water sources from pollutants, such as stormwater runoff. | 1 | 2 | 3 | 4 | 5 | 9 |
| D) | How well your water utility educates people in the area where you live about ways to protect and preserve sources of drinking water. | 1 | 2 | 3 | 4 | 5 | 9 |
| E) | Overall, how satisfied are you with your water supplier? | 1 | 2 | 3 | 4 | 5 | 9 |

## *PERCEIVED VALUE OF ENVIRONMENTAL LEADERSHIP*

7. **If your water supplier were recognized as a leader in protecting and preserving sources of drinking water would you be willing to pay a little more for your drinking water?**
   ____(1) Yes
   ____(2) No
   ____(9) Don't know

   7a. **IF YES, how much extra, in dollars, would you be willing to pay per month to cover the costs of these activities?**

   $_____ per month extra

## *SUPPORT FOR VARIOUS ENVIRONMENTAL LEADERSHIP INITIATIVES*

8. **Using a 5-point scale where 1 means "Not Supportive At All" and 5 means "Very Supportive," please indicate how supportive you would be of having your water utility do each of the following:**

|   | Not Supportive At All | Not Supp | Neutral | Very Supp | Very Supportive | Don't Know |
|---|---|---|---|---|---|---|

A) Investing in water treatment equipment to ensure that the quality of drinking water in your community exceeds, rather than just meets, Federal safety standards for drinking water ................1............2............3............4..........5............9

B) Working with other governmental organizations to restrict development and recreational activity near lakes and other water ways that are used as sources of drinking water. ................1............2............3............4..........5............9

C) Spending money to educate people about ways about ways to conserve water resources in your area ................1............2............3............4..........5............9

C) Spending money to educate people about ways about ways to protect water resources from pollution ................1............2............3............4..........5............9

D) Spending money to educate people in your community about ways to help protect the environment in general ................1............2............3............4..........5............9

E) Encouraging local governments to adopt codes that require developers to install water saving features, such as low pressure showers and low flow toilets in new developments, to help conserve water use in your community ................1............2............3............4..........5............9

F) Restricting the times and days that residents can water their lawns or fill swimming pools. ................1............2............3............4..........5............9

9. How supportive would you be of giving your water utility the authority to restrict development and recreational activity near lakes and other water ways that are used as sources of drinking water?
   ____(1) Very supportive
   ____(2) Somewhat supportive
   ____(3) Neutral
   ____(4) Not supportive

10. Overall, how well informed do you think you are you <u>about environmental issues</u>?
    ____(1) Very informed
    ____(2) Somewhat informed
    ____(3) Neutral
    ____(4) Not informed

11. Overall, how important do you think it is for your water utility to be an Environmental Leader?
    ____(1) Very important
    ____(2) Somewhat important
    ____(3) Not sure
    ____(4) Not important

12. In your opinion, which of the following best describes the overall attitude of the community where you live toward the environment? Would you say your community is generally:
    ____(1) Very willing to adopt new ways to protect the environment
    ____(2) Somewhat willing to adopt new ways to protect the environment
    ____(3) Somewhat unwilling to adopt new ways to protect the environment
    ____(4) Very unwilling to adopt new ways to protect the environment

13. Do you own or rent your home?    ____(1) own    ____(2) rent

14. How would you describe your race/ethnicity (please check all that apply)?
    ____(1) Hispanic                ____(4) Black/African American
    ____(2) White (not Hispanic)    ____(5) Asian/Pacific Islander
    ____(3) American Indian/Eskimo  ____(6) Other: _____

16. How many persons in your household (counting yourself) are?
    Under 5 years  _____    20 - 24 years _____    55-64 years  _____
    5 - 9 years    _____    25 - 34 years _____    65+ years    _____
    10 - 14 years  _____    35 - 44 years _____
    15 - 19 years  _____    45 - 54 years _____

**17. What is your total annual household income?**
   ____(1) Under $25,000
   ____(2) $25,000 to $49,999
   ____(3) $50,000 to $74,999
   ____(4) $75,000 to $99,999
   ____(5) $100,000 or more

**18. Your gender:**   ____(1) Male   ____(2) Female

**19. Which of the following best describes the community where you live:**
   ____(1) A large urban city
   ____(2) A suburb of a major city
   ____(3) A city outside a metropolitan area
   ____(4) Do not live inside a city

**THANKS FOR YOUR TIME; THIS CONCLUDES THE SURVEY.**

# REFERENCES

Coburn, John. 2000. Why integrated watershed management? In R. Coats (Ed.) *Proceedings of the Eighth Biennial Watershed Management Council Conference on Managing Watersheds in the New Century.*

Cole, Terry. 2001. Educating the public about watershed management. Water Environment Federation Technical Paper. October.

Ehlers, Laura, Max Pfeiffer, and Charles O'Melia. 2000. Making watershed management work. *Environmental Science and Technology*, Vol. 34, No. 21 (Nov.). Pp. 464A-471A.

Esqueda, Tommy, Joe Stowe, and Britt Stoddard. 1999. Implementing a regional water and sewer utility system – Wake County, NC, water/sewer master plan. Paper Presented at Water Environment Federation and American Water Works Association Joint Management Conference.

Katz, Sara. 2003. Leadership through communication. *Journal AWWA*, Vol. 95, (1). Pp. 22-23.

Laurent, Bud. 2000. All restoration is local: The critical role of the watershed resident. In R. Coats (Ed.) *Proceedings of the Eighth Biennial Watershed Management Council Conference on Managing Watersheds in the New Century.*

Lovejoy, Stephen, John Lee, and Bernie Engel. 2000. Managing watersheds: Improving the decisions with science and values. *Journal of Soil and Water Conservation*, Vol. 55, No. 4. Pp. 434-436.

Macpherson, Linda, Paul Eckley, and Francis Kessler. 2000. Public education and involvement tools to advance sustainable watershed protection. Presented at the Water Environment Federation Specialty Conference On Watershed Management. July.

Macrina, JoAnn, Walid Hatoum, and Terry Finch. (1998). Ecosystem planning for Alligator Creek Watershed. Paper presented to the ASCE Wetlands Engineering River Restoration Conference.

McGinnnis, Michael, John Woolley, and John Gamman. (1999). Bioregional conflict resolution: Rebuilding community in watershed planning and organizing. *Environmental Management*, Vol. 24, No. 1. Pp. 1-12.

Mullen, Michael and Bruce Allison. 1999. Stakeholder involvement and social capital: keys to watershed management success in Alabama. *Journal of the American Water Resources Association*, Vol. 35, No. 3 (June). Pp. 655-662

Pelley, Janet. (1997). Watershed management approach gains with states. *Environmental Science and Technology*, Vol. 31, No. 7 (July). Pp. 322A-323A.

Pendleton, Dennis. 2000. Integrating watershed initiatives with local politics: A brief introduction. In R. Coats (Ed.) *Proceedings of the Eighth Biennial Watershed Management Council Conference on Managing Watersheds in the New Century.*

Smith, Kendra. (1996) Fanno Creek watershed planning and enhancement. Proceedings of the 1996 Engineering Foundation Conference.

Spiesman, Anne, Seema Bhat, Libby Lawson, and George Rizzo. 2002. Manager to manager – Using the CCR to talk to your public. *Journal American Water Works Association*, Vol. 94, No. 7 (July). Pp. 28-31.

Tatham, Elaine, Christopher Tatham and Jane Mobley. 2004. *Customer Attitudes, Behavior, and the Impact of Communication Efforts.* Denver, Colo: AWWARF and AWWA.

United States Environmental Protection Agency. 1999. *Protecting Sources of Drinking Water: Selected Case Studies in Watershed Management.*

Wagner, Edward, Ronald Ott, and James Dunn. 1997. Watershed management addresses area water quality issues. *Pollution Engineering*, Vol. 29, No. 1 (Jan.). Pp. 82-86.

# ABBREVIATIONS

| | |
|---|---|
| AWWA | American Water Works Association |
| AwwaRF | Awwa Research Foundation |
| CCR | Consumer Confidence Report |
| EPA | U.S. Environmental Protection Agency |
| Etc | Etcetera |
| U.S. | United States |